Daniel J. Watson

HITCHHIKE TO HEAVEN

Printed in Canada.

Word Alive Press
131 Cordite Road, Winnipeg, MB R3W 1S1
www.wordalivepress.ca

Library and Archives Canada Cataloguing in Publication

Watson, Daniel J., 1942-
Hitchhike to heaven / Daniel J. Watson.

Includes bibliographical references.
ISBN 978-1-77069-089-9

1. Watson, Daniel J., 1942- --Religion.
2. Farmers--Manitoba--Biography. 3. Wives--Death.
4. Faith. I. Title.

BV4909.W28 2010 248.8'6092 C2010-905236-6

Two special ladies played a big part in my life.
To my childhood sweetheart, Margo,
who became my wife for forty-three years
before cancer took her from me.

To my partner, Frances Pettit, who shared
two wonderful years with me before
the devil cancer carried her away.

ACKNOWLEDGEMENTS

Thanks to my niece, Sheila Cameron who did a wonderful job of editing my manuscript while looking after two little kids and numerous other projects. It must have been a huge shock to her when I presented her with enough material for two books but she took it in stride and got through it.

Thanks to my sister, Ruth who used her diaries to keep an accurate account of all the dates in this book. She has been with me through the best years and the worst and I will always be totally grateful to her.

TABLE OF CONTENTS

FOREWORD

In late winter 2005, my dear wife, Margo, was in the midst of a long battle with terminal cancer. I'd fallen into a daily habit of sending out emails in the form of Margo Updates to a large group of family members and friends. Every day we eagerly read replies filled with love from our wonderful support group.

As Margo's health deteriorated she was forced to spend her last few months in hospital. Every morning I printed my Margo Updates and the replies, and brought them to the hospital for Margo to read. They were placed in her Happiness Binder and we spent countless happy hours reading these messages. The material soon outgrew the first binder and spilled over into a second.

Sometime during that period we agreed I should write a book about our experiences with cancer in hopes we could give support and comfort to others going through the same battle. We didn't want to talk only about cancer so I included some tales about my early years on the farm and my first meeting with Margo when I was barely sixteen. It is my life story to this point and includes our whole life together as we raised our family, sometimes under very adverse conditions.

The story tells how I lost faith in God and Heaven at an early age and returned to faith many years later. It also tells how Margo finally became a believer and how that brought joy and peace into our hearts several months before she passed away. She lost her fight with cancer here on earth but we both knew and accepted that she was going on to much bigger and better experiences.

In later chapters much of the story is told through actual journal entries and emails, and those pieces have been edited, mainly cut in size, in order to reduce repetition and to make for easier reading. All journal and email entries are preceded and followed by quotation marks.

I've tried to be honest in telling about some things I've done that I'm certainly not proud of. It felt important to me that they were included. The book contains many happy memories and is intended as a final tribute to Margo.

ONE
The Early Years

"Watch it, Dan. You're drifting off course again. You have to stay ahead of the plane rather than trying to play catch-up after it's too late." Ken Moore, a man barely half my age and size is seated beside me. Normally, he is totally relaxed with his arms folded across his chest as I struggle to follow his patient instructions to the letter, but two previous attempts at brief touch and go landings have his nerves stretched to the breaking point. Out of the corner of my eye, I see his hands are now resting on his knees within easy reach of the dual control yoke. If this landing is as bad as the last two, there is a very good chance he'll take over and turn my bumbling attempt into at least a controlled crash from which we can walk away.

This flying game was turning out to be a whole lot harder than I imagined when I signed up to be trained as a private pilot. Maybe I was too old to be learning a young man's game but flying a plane had been a life-long dream. I was determined to get through this course one way or another. I was forty-one and heck, that is barely middle-aged. There was no reason I couldn't do this and get my license if I could only master this stupid landing procedure.

An angry gust of wind shakes the small plane as I struggle to keep the nose lined up with that tiny grass runway. From a few miles back, at an altitude of one thousand feet above ground level, the runway had looked like a band-aid on an elephant's back end but now, a mile out and closing fast, that little strip is at least half a mile long and eighty feet wide. All I have to do is stay relaxed and try to remember my instructor's words. This is going to be the best landing that confident young man has seen in his whole instructing career.

Now, what in heck is that sequence again? Is it power, attitude, trim, or is it attitude, power, trim? Oh, man, things are happening too quickly here again but I'd best do something fast or we'll be over-shooting the runway.

I pull back on the throttle and hear the immediate drop in engine revolutions and sense the propeller slowing down. How that little engine and a thin strip of twisted metal holds this flying machine in the air is beyond me, but there is no time to dwell on that now. The nose of the plane is dropping already without that extra power up front.

Ken hasn't yelled at me yet so I must've done something right. Okay, what's the next step? Oh, yeah, attitude, and I know he's not talking about my own attitude even if he's been having serious doubts about my abilities. I pull back slightly on the control yoke and feel the nose lift and the air speed drop as we enter that peaceful glide down to land. This is the part I like best about flying. No noisy motor drumming in my ears—only the rush of dense, moist air over the shiny metal surface of this small Cessna 152.

My hand reaches the trim wheel in the dash and spins it, giving it three half-turns downward. The response is immediate as the nose pitches down, giving a wonderful view of the air strip. I watch the tiny hand on the ASI wind down until it reaches the eighty miles per hour mark and then I touch the toggle switch which operates the electric controls for the flaps. Oh wow—that is so cool! The plane slows even more as I pull back on the throttle again and amazingly, the plane is gliding down almost silently and staying right on course.

Oh, oh, I knew this was too good to last. That strong south-west wind is causing the plane to drift off course again. There is no need to worry because I know what to do. A little turn on the control yoke has the right wing dipping into the wind, and some extra pressure on the left rudder pedal straightens the whole plane out.

I watch the beginning of the runway disappear beneath me and now the ground is rushing up to meet us. Everything is working out fine and I find myself smiling as I add throttle to increase power just before the wheels make contact with the runway.

My wonderful plans go down the drain as the plane settles down much too quickly before immediately bouncing back into the air. I push the yoke ahead, the nose drops and the plane comes down with a crash that sends it rebounding skyward once more. What is even worse is that this flying machine has suddenly developed a mind of its own. It is sheering off the runway at seventy miles per hour. I become aware of Ken screaming instructions at me—something about straightening it out.

We narrowly miss a wooden runway marker and suddenly we are bouncing along in the tall grass. I remember to use the rudder pedals to steer, make a gradual left turn, and come back between the markers. Ken's voice which is normally well-modulated is coming out high and squeaky as he instructs me to stop right there in the middle of the runway. I sneak a glance at him and see that even though his hands are still firmly planted on his knees, the backs of his knuckles are white.

We sit in silence for a couple minutes as the propeller slowly ticks over in front of our eyes. I imagine Ken is wondering whatever possessed him to become a flight instructor when there are so many options open for a strong, smart, young man. I am doing some serious thinking about this crazy ambition that has caused me to spend my hard-earned dollars trying to become a pilot.

Ken finally stirs. His voice is back to normal as he suggests we do some other manoeuvres, diplomatically suggesting that there is too much cross wind for these tricky touch and go landings today.

I turn the plane around and taxi back to the far end of the strip, then head back into the wind and add full throttle. That darn cross wind is trying to push us off the runway again but in a moment we are free of the ground and climbing fast. Why is it so easy to take off but so hard to get back on the ground?

As we clear the town and head south, Ken asks if I'd like to see what it's like to fly in zero visibility. I quickly agree to anything, hoping for a break from the stressful past half hour.

Three thousand feet above ground level, we approach a solid layer of clouds and Ken takes control. We continue to climb and a few seconds later we are immersed in a dark, heavy fog so thick I can scarcely see the wing tips. Ken calmly watches the instruments and makes several turns to the left and right before asking me to tell him which way we are going without checking the compass. I guess south and he shows me that we are actually heading north.

At six thousand feet, there is a lightening in the fog above us and we start to get glimpses of blue sky. Suddenly we are in the clear, climbing out into an enormous blue and gold valley between high mountains of white, fluffy clouds. This is totally beautiful. The sun is shining hot on my face as I gaze up at the top of the clouds still several thousand feet above.

Ken gives me control and tells me to continue climbing with a gentle left turn to avoid a massive cloud directly in front of us. The plane struggles for altitude until we are caught in an extremely strong updraft that sends the altimeter spinning and my heart racing as I feel my body pressing down into the seat. The turbulence is vicious and I struggle to keep the wings level as we go shooting up past the clouds toward the blue sky and bright sun. The view is so exhilarating; I feel nothing but awe at the beauty that surrounds me.

I feel a sudden, hot rush of tears to my eyes as I whisper quietly, "Wow, this is just like hitching a ride all the way to Heaven." These words have barely become more than a thought, when I say out loud, "That's if there really is such a place."

Little did I know the direction my life would take and the wonderful things I would experience, although it would take many long years of inner struggle and denial before I would find the truth. Here is my story.

I was born on May 4, 1942, in a small hospital in the town of Eriksdale, Manitoba, the sixth child and second son of George and Dorothy Watson. My parents owned a small farm about fourteen miles from town, on the eastern shore of Lake Manitoba. There were no real roads between our farm and the town and during spring, the trails were flooded from melted snow. Dad and Mum had been through this game before and they didn't want Mum giving birth in the back seat of our Model A Ford car. Arrangements were made to have Mum spend her last couple weeks of pregnancy at her parents' house just two miles west of town. My Uncle George was in charge of driving his sister to the hospital as soon as her labour started.

On our farm, life was proceeding as normally as possible with Dad milking cows, feeding pigs and probably working some land with the horses to ready it for seeding. The older kids had plenty of chores to do every morning before heading three miles to school. My brother, Bob, was anxiously waiting for any news of my birth and was hoping for a boy. With four sisters and no brothers, he must have been feeling very much like a lone wolf.

My uncle, Will Watson, who was my Dad's oldest brother, happened to be in town the day I was born and was first to hear the news. He sent his son Douglas to the school to tell my siblings. I'm told that when school let out that afternoon, Bob ran all the way home and straight down to the garden where Dad was working, calling, "It's a boy! I have a baby brother!"

My earliest memory takes me back to when my uncle and aunt, Lister and Lily Watson, had come for a rare visit with my folks. They brought with them their four children; a big boy David, one-year old Jim, Lorna, fifteen months older than Jim, and Maisie, almost five years old.

My sister Ruth was two years older than I and we were very close. For some reason we didn't hit it off with Maisie and we left the house to get away from her. As soon as we left the house, she promptly locked the door behind us. All four doors on our house had locks that were operated by pushing a small latch on the inside of the door. We made a dash for the next door but that little brat had beaten us to it and that door too, was securely locked. To make matters worse, she was tapping on the window and making rude faces at us. In a panic, we rushed to the west side of the house to try the two doors there but she hadn't missed a trick. Her laughing face kept appearing in the windows as we circled the house.

We finally gave up and walked sadly hand in hand toward the barnyard. We crawled in under Ruth's little spruce tree where we sat with our backs pressed against the rough bark of the trunk. It was a cold day late in fall, but I only remember the red-hot anger I felt toward my cousin. I don't know how much talking I was doing at that age but I do know that the words "bad girl" were mentioned several times in the conversation.

When I related this story to Mum about fifteen years later, she did a little calculating in her head, before she exclaimed, "Dan, do you realize you were only a year and a half old? You must be kidding me." I assured her I could remember that day quite clearly and many other early memories as well.

More of my early memories had their roots in our little Anglican church in the All Saints Cemetery in Scotch Bay. Every second Sunday in the summer, and several times during the cold winter, the faithful gathered at the church to take part in the service. Sometimes we had a young student minister and sometimes the services were conducted by one or more Bishop's Messengers. These faithful young women traveled our rough country roads doing their best to spread the word of God throughout the region. As far back as I can remember, I recall Mum singing hymns and telling us the story of the Christ child who was born in a manger at Bethlehem. We heard the story so many times, especially at Christmas, that there was never any doubt in my mind the story could be anything but true.

My mother was a firm believer in Christ and she lived her life according to God's plan, being good to her neighbours and studying the Bible by dim lantern light during long winter evenings. Dad wasn't really a believer, at least back then, but he loved his wife enough that he always took Sunday afternoons off to accompany her to church. He kept us mischievous kids in line with a stern glance or a sharp elbow.

At some point during my life, I developed a strong belief that someone or something special is looking after me. I like to think I have a special guardian angel watching over me. I feel this good spirit has saved me many times from serious injury or death. When the time comes for me to leave this life, I know my good angel will be with me every step of the way. There is a saying that God looks after fools and little children. This has been proven over and over again in my case.

On a pleasant summer Sunday in 1945, Mum packed a picnic lunch and we drove up to the sandy beach at Chief's Point on the Dog Creek Indian Reservation after church. That spring Mum had given birth to her seventh and last child, another girl who my parents named Margaret. Mum was tending the baby while visiting with some lady friends in the cool shade of the huge elm trees a short distance back from the shoreline. I was playing along the water's edge running in and out of the waves when I ventured out a little too far and a huge wave crashed in, knocking me off my feet and tumbling me over and over under the water. One second I could see the sandy lake bottom, the next I was looking up through water at blue sky and bright sunlight filtered through a froth of shining bubbles. I tried to scream for help but water was filling my lungs and there was a strange roaring in my ears.

My nightmare ended as suddenly as it began when another wave dumped me coughing and spitting into shallow water near the shore. If the wave had carried me into deeper water, I would've surely drowned. I'm sure my angel was with me and gave me a gentle push in the right direction.

Life on the farm in a loving atmosphere was as good as it could get with big sisters to spoil me and a little sister to torment. The summers were wonderful, running barefoot and as free as wild animals. There was plenty of wholesome food on the table and the occasional sweet treat like the special Auntie Esther pudding made from raisins floating in a sweet juice and covered with a batter that was baked to golden brown in the wood-burning kitchen stove. To this day, I just have to think about that pudding and my mouth starts to water.

As my siblings and I grew older, our parents modeled their morals in the way they lived every day and by the way they treated their neighbours and friends. Our nearest neighbour was Neil Blue, who lived on crown land not far from the shores of Lake Manitoba. He was a small, soft-spoken man and he seemed very old to me, though in the fall of 1947, he was only fifty-nine. One day, I rode with Dad by horse and buggy to visit Neil. When Dad introduced me to Mr. Blue, the old man gravely shook my hand and told me he was very pleased to meet me. Inside his tiny, cluttered shack, the men started to visit and Neil put the kettle on the wood stove to boil. Soon we were eating a lunch consisting of tea, a small loaf of very doughy, home-made bread and plum jam from a tin can. As we ate, Neil asked me, "How do you like it, young fellow?" I replied, "It's very good, but this bread sure sticks in my teeth." This innocent remark brought a sharp nudge in my ribs from a large elbow and a stern look from Dad. On the way home, he explained to me that it had been very impolite to criticize Neil's bread when he had tried to be a good host by sharing with us what little food he had.

The fall of 1948 found me eager to attend school in the little Scotch Bay school house which was located about three miles north-west of our farm. I had been quite unhappy at being left behind as my older sisters spent day after long day away from home. The best part of going to school was getting to see my favourite cousin Georgina every day. She was born only thirteen days after me and we had been the very best of friends from the first time we met. We were baptized together on July 12, 1942. We attended school together for eight years, sometimes sitting at the same desk. The healthy competition we shared did a lot to further our education.

Within two years after I started school, our family went through a very bad time when Bob suffered a nervous breakdown. He was experiencing the manic phase of bipolar illness which had my parents fearing for his life. Dad and Mum finally had Bob committed to the Mental Hospital in Selkirk, Manitoba. Bob was tricked into going when he was told help was needed to place sandbags to stop flood waters in Winnipeg. When he realized what had happened, he was very upset and refused to see his family for quite some time. I do remember riding in the bus with Mum to visit with him once. I wasn't allowed to see Bob and was left in a waiting room while Mum visited. As we were leaving, Mum told me to look up and I saw my dear, big brother waving at me from a window several stories up. I think I cried all the way home on that long, sad bus ride.

One hot summer day when I was ten, my little sister Margie and I were weeding in the garden. From a very young age, we were taught to pull our weight and do our chores properly but we didn't always obey the rules. We had picked a bunch of fresh peas and carrots and left the garden to eat our booty. Then we started fooling around and decided to play a game we had recently invented—Blind Woman and the Sex Maniac. We weren't sure what a sex maniac was and what sort of things he might do but we were pretty sure that it wasn't good.

The rules of the game were simple. The blind lady closed his or her eyes and then took off running at top speed to avoid being caught by the sex maniac who had the distinct advantage of being able to see where they were going. It was my turn to be the blind lady and I was tearing across the pasture with my eyes closed and arms stretched out when I ran smack into the barbed wire fence at the edge of the garden. I rebounded off the fence and landed on my back with blood pouring out of my scrawny chest. It looked like I might bleed to death right there but soon the blood clotted and slowed to a trickle. I crawled through the garden fence and lay quietly in the tall weeds as Margie weeded her row and mine. I milked the situation for all I was worth and occasionally let out horrible heart-wrenching groans. Margie tried to get me to go to the house but I said there was no way I could leave the garden until my row was weeded or I would have to answer to Dad. Every time the flow of blood threatened to stop completely, I used my dirty finger nail to scratch the wound a bit in case Margie decided to check on me. After she finished weeding the allotted rows, we went to the house where I received first aid from Mum. The wound became infected and I was left with a dandy permanent scar in the shape of a small airplane.

The summer that I turned twelve, my parents decided it was time for more advanced spiritual education. Ruth and I were given a Bible study course to prepare us for the confirmation ceremony in the fall. Every Sunday, one of the Bishop's Messengers would drive out from the city for the morning church service and then spend the afternoon teaching us the word of God. We learned a lot about the Bible that summer and also were given several talks about sex education. I don't know who was more embarrassed by these talks, the children or the poor young teacher.

The confirmation took place in our little church on the evening of October 17, 1954 by the light of several gas lanterns hung from the ceiling. The Reverend Dawson held the service that night. I remember how he placed a small neighbour boy on his shoulders and pretended he was an acrobat walking across a gorge on a high wire. At one point he stood there swaying back and forth and acted as though the boy was choking the life out of him because of his fear of falling. He told him to have faith in him and they would make it across safely. After he reached the other side of the gorge, he set the boy down and told us that it is the same way with every one of us. If we only put our faith in God, He will carry us safely through all our troubles. That demonstration formed a life-long impression in my mind.

The following fall I received a rather harsh lesson at the hands of my father. One day while our parents were in town, my greedy nature caused me to become very rough with Ruth over, of all things, a batch of home-made fudge. In order to get more than my share, I man-handled her to the extent that she received heavy bruises on her neck. When our father got home, he immediately took a piece of heavy board and gave me a beating that I never forgot. I knew I deserved to be punished but I've never felt that any kid deserved to be beaten like that. Dad loved me with all his heart and I loved him too but I resolved never to forgive him for what he did to me that day and I kept that promise for more years than I like to remember.

In the spring of 1956, I turned fourteen years old and was faced with a decision regarding school. I could continue going to Scotch Bay School and take correspondence lessons with little or no help from the teacher. I could take correspondence at home, which would certainly not be easy. Another option was to board with a family in the town of Eriksdale and go to high school there. I knew very few town kids and was intimidated by the thought of living in town so I chose the fourth option, which was to quit school. In June, I graduated from grade eight with honours and that was the end of my schooling for the next ten years.

It was about that time when my dad came up with an idea which he thought might give me more interest in the farm. He suggested that instead of paying me wages, he would buy a purebred heifer and after I worked for him for one year, she would be mine. Dad was gung-ho about raising purebred cattle so with that in mind we decided to check out a small herd one mile south of Eriksdale. We spotted a beautiful two-year-old heifer with a bull calf which was only a few days old. The heifer had been accidentally bred as a yearling but had managed to produce the calf and was now a proud, young mother. One thing led to another and we purchased the pair that day. The cow's name was Blanch and I named the calf Buster. Dad's plan worked to perfection. Now that I owned cattle I became much more interested in the welfare of the farm. I started dreaming about taking over the place when the time came for my parents to retire.

The summer passed quickly doing the usual jobs of haying and harvesting. The fall was spent hauling hay home and putting it in huge stacks near the barn. This job had become much easier, thanks to a fabulous new invention called a Farmhand. It was a front end loader that fit on a tractor and used the power of hydraulics and a grapple fork to lift immense loads with the greatest of ease.

Tuesday, November 20, 1956 is permanently etched in my memory. It was about 30 below Fahrenheit but there was no wind, so Dad decided to go for a load of hay. About half a mile from home we were busy repairing an oil leak on the loader when I was accidentally speared in the buttock by one of the huge curved grapple fork teeth. Blood was gushing through the holes in my long underwear and overalls, making a crazy pattern in the snow. Pulling the mitt off my right hand, I tried to stop the flow of blood by putting pressure on the wound and I could feel the hot blood pumping through my fingers and running down my leg. Alarmed at how much blood I'd lost, my parents took me to the hospital to have Dr. Paulson check me out. Doc went about the examination in his usual calm, quiet manner. First he cleaned the wound in my butt with a clear liquid that burned like crazy as it disinfected. Next, he carefully inserted a thin plastic probe which caused a fresh burst of pain. Pulling it out, he looked at it and said, "Hmm, only four inches deep." He used two stitches to close the wound and sent me home to recover.

The next couple days were a nightmare. With no pain killers, the hurt washed over me in waves, especially when I turned over in bed. By the third day, my youthful body was starting to heal and I was able to stand up with Mum's help. Dad made a crutch for me that I used for several weeks.

Another year passed and Dad and I decided to take several days off in the fall of 1957 to go fishing at Fairford River, one of our favourite angling spots. We had a traveling partner, Ken Rutherford. Ever since Dad took up the sport of angling, we often camped with Ken. My favourite part was the evenings spent around the campfire listening to those two old friends telling one story after another.

Near the end of our trip, we decided to drive ten miles to the bridge on the Lower Fairford River with the hope of getting our limits of pickerel. By now there was very little in the grub box except for bread, butter, jam and a few fish fillets. Once at the river, we noticed a huge field with long rows of potatoes and onions almost ready for harvest. At once I remarked that we would be having fried potatoes and onions for supper and Ken happily agreed with me. Our excitement ended when Dad told us in no uncertain terms that those potatoes belonged to someone else, but the fact that Ken was in agreement with my idea gave me courage to carry out my original plan.

After dark, I went to the field and dug up one hill that was loaded with huge, red potatoes. I gathered enough for a good feed and then pulled up a couple of onions before heading back to camp. It wasn't long before there was good food sizzling in the cast-iron skillet. Ken and I dove in with a hearty appetite but Dad sat drinking tea and eating bread with butter. He never said a word but the disappointed look in his eyes made me feel so ashamed that the food stuck in my throat until I could hardly swallow. We finished the meal in silence and that night there was none of the usual happy chatter around the campfire. We went to bed early and the next morning we headed home. I was left with an unhappy memory and a good lesson in honesty.

The winter passed and a new spring arrived. I had been working hard all winter cutting firewood as well as pitching hay and doing chores on the farm. At times I regretted quitting school, especially when I was noticing every pretty girl I happened to meet, but my debt for my first cow had been paid off and I'd been earning wages for a year. My bank account was growing. My sixteenth birthday and the greatly coveted driver's license were drawing near and I was saving for my first car.

The hard work was filling out my skinny six-foot frame with solid muscle. If I looked closely in the mirror, and I must confess that I did that often, I could see the faintest trace of peach fuzz on my chin. I was becoming a man with all the strange emotions and excitement that goes with that special time. My life was soon to take on a whole new dimension.

TWO
Meeting Margo

The spring of 1958 arrived in a rush with warm winds melting the heavy banks of snow and the cries of wild geese carrying through the night. I was becoming a young man and my hormones were raging. Most of my waking thoughts were centered on the opposite sex and how I might get to meet and know some of them better. I was eagerly awaiting the arrival of my sixteenth birthday when I would be eligible to acquire my driver's license. Living so far from town without a car or license meant I had no way to travel to dances and other functions in nearby towns.

My sister Rose owned a very good car, a 1949 Chevy sedan, which she and Ruth used to travel back and forth from the farm to their jobs in the city. The highlight of my life at that stage was the weekends when my sisters would arrive for visits. They often brought their girl friends with them and some of those gals were really cute. More often than not, my sisters would pick me up and take me to dances with them. One weekend I was able to steal a few kisses from one of the visiting girls and that was an experience that left me yearning for more just like it.

May 4, 1958 finally arrived and a few days later I was the proud owner of a piece of paper that supposedly would be my key to freedom and happiness. It was a key alright but it wasn't the key to my dad's car. No amount of pleading would make him change his mind. He wasn't going to trust me behind the wheel of his precious, 1950 two-tone Ford sedan.

I couldn't believe what was happening. I was working hard on the farm, doing everything asked of me, but still wasn't allowed to use the family car which I needed to have any sort of normal social life. My parents had made it clear they didn't want me to buy a car until I had owned a license for a year so I could gain some experience. How in the world was I expected to get experience when I didn't have a car, let alone a steering wheel?

On the sixth of July, Dad and Mum left for a holiday to the west coast. I was in charge of the farm and did the chores with the help of my oldest sister Dot who was visiting from her home in Tennessee. She was a wonderful cook and made sure we never went hungry. With my parents away I started thinking about obtaining my very own set of wheels. I asked Rose to look around in the city for a suitable car.

Within a week or so, I got a call from Rose saying that a fellow in the rooming house where she lived had a car for sale, a 1941 Nash sedan. The fellow told her the car was in good running condition. The price was certainly reasonable at fifty bucks with its present tires or sixty bucks if he put better tires on it. I immediately told her to pay the higher price so I wouldn't be bothered with flat tires while out on a date. Oh, the innocence of youth and how little did I know about dealing with people in the real world.

That Friday, July 18, I waited impatiently to get my first glimpse of that car, the wonderful chariot that was going to transform me from shy farm boy to handsome prince. Rose brought the car out after she got off work, with Ruth following in the Chevy. The days are long at that time of year but there was precious little daylight left when they finally pulled in.

At first glance, the car was all that I had hoped for. It was only a few moments before I was behind the wheel taking it up the road for a spin. The headlights didn't emit much more light than a bright candle and a closer look had revealed this model didn't have the fold-down seats I'd been expecting, the tires were bare, and the motor back-fired if you stepped on the gas too hard. None of that really mattered; I had my own set of wheels and it wouldn't be long before I knew every girl in town. The next morning Rose drove me to town to arrange for the vehicle registration.

What was even better than having a car at last was having my cousin Jim here for a visit. He was not quite sixteen years old but he had quit school and his parents had sent him from their home in Thunder Bay, Ontario to find summer work as a farm hand. Jim and I had only met each other a few times but we had been buddies from the first meeting. We were soon on our way to the town of Lundar to check things out.

It didn't take long to realize there were a whole lot of things wrong with that old car. Number one affliction was a battery that must've been at least half as old as the car. If the motor stalled, it was very hard to get it started again. By evening we'd managed to limp it back home and were thankful to be there.

We were still out tinkering with the motor in the last dim light of day when we heard a car pulling into the lane leading into our yard. To our surprise, two cars pulled up to the house, each loaded with young people who spilled out laughing and happy and free as the breeze. It was my sisters and some friends of theirs from Lundar and they were getting set for a party at the beach. We all stood around in the headlights as introductions were made. My sisters had a city friend with them and also two former Lundar girls. These girls worked in the city but still liked to come home on weekends. They were Bertha Nordal and her younger sister Margo, both riding in the other car with Margo's boyfriend, Gerald, a young man from Lundar.

There were other young people in that car as well but I don't recall who they were, because from the moment I first saw her, I couldn't take my eyes off Margo. Beautiful and vivacious, wild as the wind and as slim as a young willow tree, she was drinking beer from a bottle and laughing as if she hadn't a care in the world. When I was introduced to her, I forgot about my shyness and stepped forward and took her hand in mine. As her gaze met mine, something strange happened to my heart. I was barely sixteen years old but my fate was sealed in that magic moment. I had met the girl who would be my partner for life, though neither of us had any such notion at the time.

Jim and I piled into Rose's car with the other three girls and we drove one mile south to Long Point Beach, a favourite swimming hole for people of all ages. I could barely wait to get out of the car at the beach so I could talk to Jim about that young Nordal girl and whether he thought there was any chance I would be able to date such a lovely creature. One thing for sure was that without a car it was hopeless. My old car suddenly became extremely important.

The next morning I was full of questions for Rose as I tried to discreetly quiz her about Margo—how old she was, where she worked and where she lived. I was rather disappointed to find out she was two and a half years older than I, a real woman of the world already while I was just a snotty-nosed kid in comparison. That didn't make me any less determined though and I was sure if I got a chance to see her again, I'd be ready to take full advantage of it.

Monday morning arrived and we were out of the house early, eager to get the chores done. We wanted to try out the car and see if we could meet some girls in town. A whole new world had opened for Jim and I. Most of the girls in Lundar were eager to meet us now that we had wheels. It usually took only a turn or two up and down Main Street, idling along with one arm casually resting on the edge of the open window and girls would appear like magic. Sometimes our new lady friends would hop in for a ride and maybe a soft drink and a swim out at Lundar Beach. It was a bit embarrassing when the motor refused to start but we soon found it only took a little push from several willing hands, a quick pop of the clutch and we were back in business.

We put a lot of miles on the car that week and were plagued with flat tires on every trip. It soon became apparent that I had to invest in a hand pump and a kit to patch flat tires. We got lots of practice and soon became efficient at repairing tires even in the dark aided by the dim glow of a flashlight.

The honeymoon was over a few days later when Dad and Mum got home from their trip to find I'd gone against their wishes and purchased a car. In vain I explained to them there was no way I was going to get any experience driving if I didn't have something to drive. Dad was so angry, he refused to discuss the issue at all but Mum tried to explain that Dad was really worried I would start drinking and driving and get into an accident. I told her she had no need to worry but I had to turn my back on her to avoid the concerned look in her eyes. It seemed as if she could read my mind and knew I had already sampled my first beer while they were away.

The following Saturday, Jim, Ruth and I were at Long Point Beach where a large crowd of young people was enjoying the hot summer weather. We met up with the Nordal girls who were riding with their brother Paul. Somehow we all decided to drive up to Chief's Point. We all piled into Paul's car and I couldn't believe my luck when I found myself sitting beside Margo. She was so much fun to be with because of her happy nature and she looked and smelled great. Unlike most of the girls I knew, she wore very little make-up and no perfume but she smelled wonderful just the same, with a kind of fresh fragrance, like a bouquet of wild flowers.

At the beach we all went into the water where I was surprised to learn Margo couldn't swim a stroke and was actually afraid of the water. After cooling off a bit we ate our picnic lunch and then started for home again. Again, I found myself riding beside this captivating gal. Jim and I were vying for attention as usual by telling all sorts of naughty jokes that brought peals of laughter from our captive audience.

When we reached home, we sat there for an hour or so drinking beer and gin. I couldn't take my eyes off Margo and was even growing bold enough to casually drape my arm on the back of the seat behind her with my hand resting lightly on her shoulder. If she noticed, she made no comment. When she reached for a cigarette I was quick to reach for my lighter to light up both our smokes. As I lit her cigarette, she cupped her hand around mine to shield the flame from the balmy breeze that was blowing through the car. The gentle feel of her soft hand made my silly heart do back flips while my breath caught in my throat. I knew my world was never going to be the same.

The following week Jim and I were split up when he was hired by my Uncle Kris and Aunt Alice Thorkelson to help in the hayfields. We were busy haying too and that meant an end to the care-free existence I'd been experiencing. Jim and I still had our Saturday evenings and Sundays off and we made the most of them. We were doing some drinking and I loved the way it made me feel brave and confident, enabling me to walk up to anyone and strike up a witty conversation.

Over the next few weeks, I managed to talk to that young Nordal girl every chance I got but it was plain to see she was only being polite to me because I was the little brother of her city friends. I was bold enough to keep suggesting that perhaps she would let me pick her up at her parent's house some Friday night and we could go dancing. That was enough to bring a smile to her lips and a soft laugh which set my heart pounding. I wasn't wasting my time pining for her though because there were plenty of other girls who were interested in me now that I had my own wheels. The weeks were passing quickly and I was having the time of my life.

The last day of August found a bunch of us at the annual corn roast at Lindell's barn dance. After eating our fill of juicy corn drenched in butter, we walked up the steep stairs on the end of the barn to watch the band tuning up for the dance. At the dances in those days the girls sat around the edges of the hall and the young men stood at the back of the hall in a stag line. We would talk and rough house with each other, all the while wishing we had the nerve to ask some girl to dance. My wish came true that night when Ruth dragged me up to dance a Butterfly dance with Margo. The dance involved very little skill but lots of energy which suited me fine. It gave me a chance to put my arm around Margo's slim waist and see her face shining with laughter as we tried to swing each other off our feet. It also gave me great satisfaction to note that Gerald was too shy to dance even once with her and he stayed in the stag line all night. In spite of my jealousy of his relationship with Margo, I was fast becoming good friends with him because he was a genuinely good person who didn't have a mean bone in his body.

About mid-September, things turned rosy for me when Margo finally accepted my invitation to accompany her to a dance in Lundar. Perhaps she was just tired of refusing my frequent requests but I like to think she'd been impressed that I'd enough nerve to dance with her, even if the dance involved my sister as well.

That Saturday, I spent the afternoon washing my car, trying to make it look just right for this special date. Outwardly, the car looked great—fairly shiny with no rust spots—but the performance was poor. The motor had developed a knock and was burning almost as much oil as it was gas. One front wheel had a habit of coming loose when you least expected it and every so often you had to get out to tighten the lugs before the wheel fell right off.

I wanted to make a wonderful impression on Margo so I spent almost as much time on my own appearance. I carefully shaved the peach fuzz off my chin and bathed until I was squeaky clean. My curly hair was teased into a fantastic duck tail which was all the rage back then. I put on my black shirt which was filled with all sorts of shiny, sparkly stuff and slid my skinny legs into my tight blue jeans before stepping up to the mirror for an inspection. I wasn't altogether happy with what I saw there. Over the past two years I'd had a tremendous growth spurt. I measured six foot one inch but weighed only one hundred thirty-five pounds with elbows and knees jutting out at awkward angles, making me look like a real live stick man.

Shortly before sunset, I pulled up to the house where Margo was visiting with her parents for the weekend. Einar and Flora Nordal had raised a family of eleven children in that little house and still had the four youngest kids living with them as well as two older boys. I soon found myself standing in the tiny kitchen meeting Margo's parents for the first time. My parent's careful lessons in good manners served me well as I shook hands with the old people. We made small talk about nothing special until Margo appeared. She looked so beautiful that I could feel my face flaming. I couldn't think of a single sensible thing to say. I felt I was in a dream as we made our way out to my car.

I stepped on the starter button and that tired old motor clattered to life. I breathed a silent prayer of thanks to God for making my dreams come true and asked Him to keep the old car working well for the remainder of the evening.

Half way to town, a cloud of steam burst out from under the hood as the radiator boiled over. The water pump had a bad leak and I'd forgotten to top up the radiator before leaving home. There was nothing for it but to take off my shoes and socks and wade into the ditch to haul water in a tin can to fill up the radiator. It took several trips and by that time my feet were muddy. It took awhile before I was able to get them clean enough to put my shoes on again. All the while Margo kept up a light chatter as if this was her idea of the perfect date. Once more we started off but hadn't gone far when a tire went flat. Cursing under my breath, I quickly got my tools and changed the tire.

Things improved once we reached the dance hall. I saw her look at me in surprise when I paid her admission as well as my own. It cost me a bit of my hard-earned savings but I was sure I'd made a good impression. Once inside the hall she took my hand and led me through the stag line to an empty spot along the side of the hall. We chatted for awhile until she asked if I would like to dance. I had to admit I didn't know much about dancing but I was willing to learn if she was willing to teach me. What followed were several hours of the most intense excitement, as she struggled to teach me the steps to old time waltzes, fox trots and polkas. I stumbled all over her little toes with my big feet but that didn't seem to matter to either of us. We were having such fun and truly enjoying each other's company.

As we danced she gazed up at me and I marvelled at her exotic beauty. She had inherited her blonde hair and fair complexion from her Icelandic father and her rather thick eyebrows and dark brown eyes from her Scotch-Métis mother. I thought she was the most beautiful creature I'd ever seen. Something in the way she was looking at me suggested that perhaps she was seeing me in a different light for the first time.

After the dance, we walked out through the dark night and climbed into the car. Another miracle took place when the engine started at once. I suggested a drive out to Lundar Beach to see if any of our friends were hanging around out there. Margo immediately accepted and a minute later she was sitting close beside me in the middle of the bench seat. Things were going exactly as I'd been planning all week and once again I said a silent prayer asking that nothing go wrong to break this magic spell.

Of course, it was all too good to last. A tire went flat and I didn't have another spare. There was nothing for it but to take the tire off the rim and patch the tube, with Margo standing beside me holding the flashlight. By the time we were back in the car she had seen enough trouble to last all weekend and insisted I drive her straight home. Once there she leaned over and gave me a quick kiss that was all I could've wished for. A second later she was running for the house without a backward glance. I drove home with no further problems, my heart singing all the way. The evening hadn't gone as well as I'd hoped but it could've been worse. The best part was that we had so much fun together despite all the set-backs. I would've been so happy to continue dating Margo but by the following weekend she was back with Gerald and I was seeing other girls from the area.

Soon that old car was giving me more trouble than it was worth and the mechanical problems were getting worse all the time. It was time to face the fact that I couldn't afford to keep it running. Although it was my life-line to a social world, the car made its last sad trip, a short drive to a sheltered spot where it was left to rust and rot until it was hauled away by a scrap dealer some twenty years later. I made up my mind to get a decent job and save until I had enough money to buy a dependable car.

I purchased twenty gill nets and went into a fishing partnership on the frozen surface of Lake Manitoba. A neighbour, Fred Ross, owned a small fish camp which we parked near the shore on my Uncle Albert's land. Uncle Walter formed the third member of our partnership and we did quite well up until Christmas time. My twenty nets paid for themselves on the first lift and every lift after that brought another pay-day which I faithfully deposited in the bank. My savings account grew and I dreamed about purchasing another car. The hard work was good for me as it settled my mind and strengthened my body.

Once the ice was thick enough, we moved the camp onto the ice about half a mile from shore. I loved living there even though it was a poor camp with no insulation. Exhausted after each day's hard work, I would fall asleep to the sound of ice booming and cracking under us. It was a common occurrence to find the pail of drinking water frozen right to the bottom by morning. A few minutes later, fire would be crackling in the tin heater while porridge cooked in a pot. My two experienced companions taught me a lot about hard work. I learned how to make the most of my time by doing things properly the first time.

The new year of 1959 arrived and the fishing dropped off in our Scotch Bay area to the point where Walter and Fred decided it would be better to pull the nets out and call it a winter. Within a few days I found work with the Lee brothers, Watson and Matt, who were getting ready to move their camp about twenty miles south to a spot in the middle of the lake. Watson owned a very good, well-insulated fish camp and living there was like moving into a five-star hotel after the rough conditions in the previous camp. The work was hard but the brothers treated me well and my education continued. The weather was extremely cold. The ice grew thicker until it reached a depth where you got your hands wet while using a chisel to open a new hole.

On weekends we would drive back to Scotch Bay. The little Ferguson tractor pulled the caboose, which was loaded to the roof with boxes of fish. These fish had been buried in a snow bank all week to keep them from freezing. The fish were sold to the local buyer and I was given Sunday off to go home and rest up to get ready for another week.

By the time the fishing season closed, I had built up a fair bank account and it was time to buy that much-needed car. To ensure I spent my money wisely, I asked Wilfred Blue for assistance. Wilfred had been making a living farming and running an auto wrecking business and nobody knew cars or engines better than he did. I worked for him cutting firewood for several days and in exchange he took me to the city where we made the rounds of the car dealerships.

I was looking for something flashy with a big motor with lots of power but Wilfred was much too wise for that. It was well after dark by the time we found a car that satisfied him, a very ordinary-looking 1947 Dodge sedan with a six cylinder motor that would be easy on gas even if it was no power-house. The tires were like new, it had a good battery and there wasn't a rust spot to be seen. We haggled with the salesman about the price and within an hour we were out of there with my wallet lighter to the tune of three hundred bucks.

Soon I was running around every weekend. Every so often I saw Margo and always spoke to her, although she was still resisting my efforts to date her. Finally I must've caught her in a weak moment and she let me drive her from a dance in Lundar to a house party a few blocks away. That seemed to be the turning point. By early July we were seeing each other exclusively and totally enjoying being together. That little lady was so much fun to be with, no matter whether we were sun-bathing at the beach, having a picnic lunch, or dancing in town or at one-room school house dances. We always had beer along. A case of twenty-four bottles cost only $4.20 and four people could chip in a buck and a bit and have quite the party.

I was working for my dad to pay off the debt I owed him for another heifer he had purchased for me the previous fall. My herd of cattle was expanding and I had to do my share to put up hay for them. It must've driven my dad crazy to see me leave the field early every Friday evening in order to meet my sweetie in town and he saw precious little of me on Saturdays and Sundays. Luckily my two big sisters pitched in and took my place in the hayfields when I was running around the country.

My idyllic lifestyle came to an abrupt halt on the evening of August 1 when a moment of carelessness caused an accident which badly damaged my car and resulted in me losing my driver's license for a year. I had to phone my parents to come to town to tow my car home. I mentally braced myself for some harsh words from my dad but his words were kind and gentle. He told me he'd been worried all summer about getting a phone call to tell him I'd been killed in an accident. He was very thankful it hadn't been worse. That made me feel a little better but the fact remained my car was disabled. The worst part of the whole mess was admitting it was my own fault.

By now Margo and I were deeply attached to each other and the following weekend found us dancing in Lundar again with her ex-boyfriend, Gerald, doing the driving for us. I think we made a deal where we bought the beer and he bought the gas for his car. From then on he was a willing driver and a wonderful friend to both of us.

November 21, 1959, found Margo and me heading to the city with some friends of ours. It was Sadie Hawkins Day and Margo had asked me to go with her to that special dance where the girls did the inviting. If the guy accepted the invitation, he had to wear a silly corsage which his gal friend made from balloons, ribbons and all sorts of fancy junk. I must've been a strange sight, over six feet tall, skinny as a rail and wearing a borrowed jacket which was three sizes too small. We went to Patterson's Barn dance in the city and at first I felt completely out of place around all the strangers. It was so much fun being with my special gal, I soon relaxed and had a good time. I was becoming a fairly good dancer and that suited Margo because she loved to dance even more than she loved to drink beer.

Earlier that fall, I'd been unable to find anyone to form a fishing partnership with, so I accepted an offer to fish for my Uncle Walter. When the season opened we moved into his shack near the lake. My wages were only $90 per month plus room and board but because I had no wheels, I was able to save a fair bit of money over the winter. Margo and I continued to see a lot of each other on weekends. I would catch a ride to town with a neighbour and somehow always managed to find a way home.

By the end of January, 1960, the fishing had dropped off to the point where my uncle was barely making enough to pay the expenses. He thought it best to pull the nets and call it a season.

The previous year my cousin, Jim, had hitch hiked to Vancouver with his older brother, David. He'd written often and eloquently about the marvellous life he led there. By now Jim was home in Thunder Bay living with his parents again. I contacted him and we made plans to hitch hike out to

Vancouver where we hoped to find work. The only fly in the ointment was the fact Jim was completely broke. That didn't worry me because I was willing to pay all the expenses in exchange for his traveling experience.

I hated to leave Margo for any length of time but I knew she wouldn't be home crying her eyes out for me while I was gone. We had talked it over and agreed that we were free to date others while we were apart.

Jim and I left Winnipeg on Feb 20, on a viciously, cold winter morning and made it almost to the Saskatchewan border that evening. Over the next month we had many experiences and met many people, both good and bad. We never made our fortune and never even found a job but I treasure the memories of that trip.

We spent most of the time in Vancouver, BC, but we also spent a week in a logging camp with my brother at Red Lake, near Kamloops, BC. Shortly before we left to come home, Bob tried to convince me to stay with him until after spring break-up so we could take the bus back to Manitoba together. Without giving it too much thought I told him I was heading out on the road with Jim again. That was a decision that haunted me for most of my life.

We were home before the end of March. On May 16, our family received a shocking phone call from Kamloops. My brother had committed suicide in his cabin in the mountains. I remember well the moment I first heard that horrible news. It felt as if I'd been dumped into icy water and couldn't catch my breath. My second emotion was extreme guilt because I was ashamed of myself for not staying with him when he needed me most. I was certain that if I'd stayed until the spring thaw brought an end to his sawmill job, he would've been back in Manitoba with me. I was also ashamed of Bob and terribly angry at him for taking his own life. Those disturbing feelings would stay with me for many years.

*   *   *

Margo and I were dating but things weren't always rosy between us. That led to several short break-ups where we went our separate ways until we cooled off. We both continued to date others during our times apart. During one of our brief on-again periods, Margo and I decided to travel with our friends, Archie Davidson and Bernice Lyndstrom to Swan River to take in the weekend rodeo and fair. The other three were confirmed city slickers by then while I was just a farm boy working for my Uncle Jack and Aunt Mary Ross. Archie, Bernice, and Margo came out from the city one Friday evening after work, and stopped at the Ross farm where I was eagerly anticipating a long, dark ride smooching in the back seat with my sweetie. We didn't have much time to spare but Archie had a new, big car and he was a fast driver especially when he was drinking as we all were that night.

We got to The Narrows in time to catch Alec Freeman, the ferry operator, just before he left to go home. The ferry only operated in daylight hours. Once we were safely loaded on that tiny craft, we relaxed and enjoyed a fresh beer while kicking back as we talked about the great weekend that lay ahead of us. The roads weren't good and it was breaking daylight by the time we straggled into town. We slept in the car for awhile until the sun was up and then found a cafe that served us coffee with our breakfast.

Once the fair grounds opened, we wandered around looking at various exhibits and wasting a few dollars at those stupid booths where you were guaranteed to lose your hard-earned bucks. The rodeo cost another couple of bucks each and our funds were getting low. We barely had ten bucks between all of us by late afternoon when we wandered into a huge tent featuring a strip tease. At first it was all fun as we giggled at the sight of three or four middle-aged women strutting their stuff around the tiny stage to what was meant to be provocative music. The gals finally stripped down to little bikini type outfits and then the show stopped. The announcer came out on stage and told all the women in the audience that they would have to leave as what we were about to see was for men's eyes only. The girls stood up and looked at us meaningfully. I timidly asked Archie if he was going to stay to see the rest. His answer was an emphatic, "Hell, yes." The girls stomped out angrily while Arch and I settled back in our seats to watch the rest of the show.

Another dollar was collected from each of us and then the tired-looking gals did a few bumps and grinds before whipping off their bikini outfits to reveal equally tired-looking boobs and rather doubtful-looking butts. Even these well-worn parts of their anatomy were covered with pasties and G-strings. We were allowed to feast our eager eyes for about fifteen seconds before the lights dimmed and the gals ran off the stage. We skulked out of there and found our girls' sun-burned faces set in serious scowls as they informed us they wanted to go home immediately.

We didn't have enough money left to pay for supper and we were all starving. We settled for a couple cups of coffee to soothe our raging hangovers and hit the long road back to The Narrows. On the way north we'd been all cuddled together but now the girls were both sitting as far over in their seats as they could get. If those doors had come unlatched, they would've tumbled out on the road. We got to the lake shore on the west side of The Narrows just in time to see the lights of Alec's car fading away over the top of the hill as he headed home.

Margo wasn't talking to me and was still keeping her distance as we settled down for the long night in the car. When it got dark enough, I sidled across the seat, thinking I might snuggle up a bit while she was sleeping but I was met with a sharp elbow in my skinny ribs. I retreated to my side of the car muttering angrily under my breath that if I ever saw this stupid broad again it would be much too soon to suit me.

The night was dark and extremely hot but each time we opened the windows to let in air, hordes of ravenous mosquitoes poured in. There was nothing for it but to shut the windows, kill off the remaining mosquitoes and try to catch a bit more sleep. We'd been unwashed for a little too long and the fact that some of us were letting out sneaky, little beer farts didn't help matters any. There were lots of angry accusations being hurled about and I must admit that some of the complaints were justified. Daylight finally came and we were a sad, mosquito-bitten bunch, all still giving each other the silent treatment.

You would wonder how our relationship could survive such an awful setback but Margo and I were soon back in each other's loving arms. Unfortunately, it wasn't long before she started talking about going west to visit Bertha in Vancouver. I didn't want to see her go and tried my best to talk her out of it. She wasn't listening, and in the first week of July she was on her way. She found work washing dishes in the kitchen at City Hall but soon grew to hate that job. On August 9, she hopped on a bus and rode back to Manitoba.

Margo and I continued dating that summer and fall, and by winter I was back working for Uncle Walter on the frozen lake. The drinking and partying continued every weekend that Margo could make it out from the city. It is truly a miracle I survived that winter the way things were going. I had my driver's license back and had traded in my faithful car for a beautiful 1953 two-door Meteor. It had a V-8 motor with enough power to spin the back wheels with just a slight push on the gas pedal. A typical Friday evening found me rushing through my last few chores at the fish camp and then driving madly down to Lundar to meet up with Margo. We went dancing and then to house parties. Usually by three in the morning we would be parked in Margo's parent's yard.

Sometimes we would go into the kitchen to see what we could find to eat and most times her mum would climb out of bed and join us, her round face smiling with joy to see us safely home. She would dig out a loaf of wonderful homemade bread and a roll of rullupylsa which was an Icelandic delicacy—mutton flanks rolled up with spices and boiled until tender.

More often we would just sit in the car necking, talking and drinking beer if we had any left by that time of night. We were finding it increasingly hard to be apart from each other and often I would stay until I barely had time to get back to the fish camp for another day's hard work. I fought sleep on that twenty-mile drive but always made it safely. I'd tip-toe into the shack to quietly undress and slide under the covers. I'd just be falling into a blissful sleep when the silence would be broken by the shrill sound of Uncle Watt's alarm clock. A minute later he'd be lighting the fire in the stove with the usual dose of coal oil to get things started off with a bang. Two minutes later he'd call me to get the supplies into the caboose.

Saturday would go by quickly as I worked hard at my job and in no time at all, I was back at Nordal's house to pick up my sweetie. She would meet me at the door bright-eyed and bushy-tailed after sleeping most of the day and we would be off on another round of dancing and parties. Sunday morning would find me sneaking into camp again just in time to go to work. I was being paid a monthly wage and was working seven days a week. On this day, time would drag as I was almost falling asleep on the job. I would barely finish my supper before falling into bed to sleep fitfully with all sorts of weird dreams brought on by my state of exhaustion.

* * *

By mid-March, 1961, I was out of a job again and by chance learned that a man from west of Gypsumville was looking for two good men to work cutting logs with chain saws. Art Nordal came with me to meet Ken Castle at his camp in the bush. Ken had built a small portable sawmill which he hoped to use to saw poplar lumber to build fish boxes for sale.

We were hired immediately and put to work that very morning cutting down large poplar trees and bucking them into eight foot logs. I thought I knew all about work and thought I was in shape but quickly found out this job was much tougher than I imagined. The snow was deep, well up over our knees, and the chain saws seemed to weigh a ton. We kept at it and as the weeks went by, the work became easier.

That job kept me busy until mid June. The summer passed quickly as I helped Dad in the hayfields and spent several weeks working for Uncle Jack bringing in his grain harvest. It was the most wonderful time in my life as Margo and I continued to see each other at every opportunity. We reached the point where we didn't want to be apart at all. I would either spend the weekend nights at her parent's house or she would stay in my sister's room at my parent's house. Neither of the houses was very large and this sleeping arrangement led to a rather humorous incident at Nordal's house one Saturday night.

We had come home from a dance and had a light snack before she tucked me into a pull-out bed in the living room. Her parents slept on a bed just across the room from me. After a last quick kiss from my sweetie, she tiptoed upstairs to her bed. I quickly fell asleep and was soon dreaming that Margo was crawling into bed with me. The dream was so real I could feel her warm body on top of me. I reached up and pulling her face down to mine, gave her a big, mushy kiss right on the lips. I heard the sound of giggling and became aware of the taste of rullupylsa on my lips. I woke up to find I was clutching Margo's brother, Kenny, who had been crawling over me to take a spot on the other side of the bed. He'd eaten his fill of rullupylsa before coming to bed, not expecting to be greeted so warmly. By the time we quit laughing, everyone in the house was awake and laughing with us.

The first day of September found me back at work in the bush camp. The work was hard and tedious as we worked long hours sawing square timbers into boards. The last week of the job crawled by as we crouched on hands and knees sorting the boards into bundles which were wrapped with wire for shipment to the city. On the last day of October we packed up the last of the boards and put things away for winter. By noon we had eaten lunch and been paid off with a generous bonus so were ready to strike off for home. Before we left, Ken pulled out a bottle of whiskey and offered us a drink for the road. Three bottles later, we shook hands with him and wished him well. That was some wild ride home and one I would just as soon forget.

I was soon fishing again, this time for Gerald and his dad, Oscar. We spent the weeks in a shack beside the Swan Creek fish hatchery. We worked hard, had fun together and the fishing was good.

I'd been quite unhappy during that period of time because Margo had gone to Vancouver again on September 21. She was back in Manitoba in time for Christmas and from then on, we were totally committed to each other. Our late-night talks were more serious as we contemplated a lifetime spent raising a family together. Time dragged so slowly during the week while we were away from each other. The biggest obstacle to overcome was to find jobs that paid better wages but we were determined to find a way to be together for good.

By the end of January, we'd decided to get married even if it meant starving to death together. Most couples get engaged when the young man picks out a beautiful diamond engagement ring and gets down on one knee and proposes unexpectedly. We did it our own way, the same way we would live the rest of our lives, as together we went to a store in the city and picked out a perfect ring. It was more than I could afford but it was the one we chose together. It was worth much more than the price I paid when I saw her hands shaking so hard, she could barely get that ring on her finger.

Once we were officially engaged we became more careful about how we spent our money. We cut way back on drinking and partying, choosing instead to spend quiet times together as we made plans for our rather uncertain future.

The spring of 1962 arrived and we were no closer to landing good jobs than we had ever been. Margo worked in a sewing factory and her take-home pay was about $28.00 per week. I was back to working for about seven bucks a day for whichever farmer would have me. With love in our hearts and the ignorance of youth providing courage, we planned our wedding for July 7. We were going to be together and nothing else really mattered.

THREE
Married Life

It was nearing summer of 1962 and Margo and I were planning our wedding for July 7. It had been almost four years since we met and we were now firmly committed to joining our lives forever. We were finding it increasingly hard to be apart all week while she worked in a sewing factory in the city and I helped my dad on the farm.

One of our first tasks was to find a preacher to marry us. Rose solved that problem when she arranged to have us meet her preacher, a tiny old man named Canon Hughes. We were surprised when we learned that we had to meet with him on several occasions as he tried his best to prepare us for married life. We must've looked like a couple of young children to that kind, elderly gentleman and I'm sure he was worried about what lay ahead of us.

The following weeks passed quickly and before we knew it, the big day arrived. As was the custom, the groom wasn't allowed to see the bride before she entered the church so I was banished to the tiny room that we had booked in the old Aberdeen Hotel in downtown Winnipeg. Why we picked that particular hotel is beyond me but I suppose it was because it was the hotel that Mum and Dad chose when they stayed overnight in the city.

I spent the morning eating green, seedless grapes which were just coming into their own and I must've eaten a couple pounds of the delicious things. When the time came to meet at the church, I was suffering from stomach cramps. For awhile I feared I might have to leave the church in the middle of the ceremony to attend to more urgent business.

It wasn't long before I was seated in a room with my brother-in-law Alf Anderson while the preacher gave me some last minute instructions. He jokingly told me that if I had any notion of taking off, I had best do it now before it was too late. I wonder if he gave that same advice to all the young men he married.

Within minutes, I was making my way on shaky legs into the church where a small congregation of immediate family members was gathered. Margo's family was on one side with her little mum smiling proudly. My family was on the other side with poor Mum looking very worried as she smiled at me through tears. There wasn't long to look at them before the organ began playing the wedding march. The little preacher man beckoned to me and I leaned forward to catch the words when he smiled and said, "It's too late to beat it, now." He was right of course, as I was ready for whatever life might bring me and my beautiful sweetheart. Perhaps it was just as well that I couldn't see the near future or maybe I would've run out of there before saying my vows.

The bridesmaid, Margo's sister Kay, entered the church and made her way slowly to the front and then came my first glimpse of my bride, totally gorgeous and radiant in her beautiful, borrowed wedding dress. It seemed like only seconds before her dad led her to me and our hands joined. The rest of the ceremony was a blur because I couldn't take my eyes from Margo's face as I marvelled that this beautiful creature could possibly love me enough to commit the rest of her life to being with me. I ignored my stomach cramps long enough to repeat my vows and take our first walk down the aisle as a married couple. The photos that were taken of us, surrounded by loving family members on the steps of the church, reflect the happiness in both of our brave young hearts that day.

A wedding feast had been prepared at Kay and Alf's apartment and a good time was had by all. It seemed to be a very long time before we were able to escape into our car and drive to our room downtown. We had to make our way into the hotel past several drunks who leered and made rude remarks but that couldn't dim our happiness. We were together and alone at last, with nobody to tell us what we could or couldn't do and we intended to make the most of it and just be happy.

The next day we drove out to spend the weekend at Clear Lake in the Riding Mountains. I suppose there were plenty of sights to see but we hardly left the cabin for our entire stay. We had brought all the groceries we needed to cook our meals and the rest of the time we just enjoyed being together.

We were back at my parent's farm early the following week trying to come up with a plan of action. I had a bit of money saved but that would be gone quickly if I couldn't find a decent job. We put an ad in a farm paper to try to obtain a good job on a farm which was the kind of work that I loved and was experienced in. In the meantime, I would help Dad get as much of the haying done as possible.

Dad and I were making good time that week but on Saturday morning, I had to tell him I would be leaving the field early. Mac Miller had two lambs he wanted me to help butcher over at his brother Kemmy's place. I liked butchering of all kinds and was very good at butchering sheep. I also loved my neighbours, the Miller families, so I was looking forward to a pleasant evening.

About six o'clock, I drove the tractor over to where Dad was busy stacking hay. I told him I was leaving to go and kill the sheep. He smiled, with a twinkle in his eye, and told me not to kill them too hard. That was a strange remark but in my excitement I never thought too much about it.

At home I gobbled down the good supper the women had waiting. Once my belly was full, I was soon urging Margo to hurry up so we could get going. Then that little gal really floored me when she said that I wasn't going anywhere until I got cleaned up and had a shave. I told her I was going to butcher sheep so there was no need to clean up. To my surprise she turned real stubborn and refused to move until I did as I was told. I went to have a bath complaining bitterly that I would get blood all over my clean clothes. She told me firmly that I could wear coveralls on top of my good clothes and take them off after the job was done.

Half an hour later we pulled into Miller's yard well before sundown. Mac was there already as well as a bunch of other friends. Some were my good drinking buddies and it wasn't long before I had a beer in my hand. I kept asking Mac where the lambs were and "Shouldn't we get busy before it gets dark?" He kept coming up with excuses and all the while more people kept arriving in the yard until at one point I remarked that we should be having a party. The surprise was on me when I was finally told that our good friends were putting on a combination party with a shower for Margo and a stag for me. The idea was to raise money to get our marriage off to a good start. As I looked around at all the smiling faces, I almost burst out crying but I held back the tears and declared that it was party time. Of course Margo and my parents were in on the plan from the first.

The women stayed upstairs for the shower while the men went down to the basement which was all decked out for the party. Dozens of cases of beer were stacked on one side of the room and it wasn't long before the party was in full swing. We sat around talking and listening to music, and if you got feeling kind of dry you'd go to the bartender and buy another beer. My guitar had been in the car and I sang my usual repertoire of naughty songs to the very appreciative audience.

The party went until the wee hours and as might be expected, most people got very drunk. The party was winding down when a couple of neighbours, Reino Kaartinen and Raymond Brown, got into a terrible fight. Ray's dad, Donald jumped into the fray and the three of them staggered sideways across the room until they all fell with a huge crash onto the pile of beer cases. There were still half a dozen full cases and a lot of bottles got broken when they hit the cement floor. Kemmy got real mad and cussed out old Donald, who happened to be his brother-in-law. For a few minutes it looked like mayhem was going to reign supreme in that little basement. Luckily, cooler heads prevailed. The tired combatants offered to pay for all the broken bottles and that seemed fair enough to me. Actually I was quite happy about the fracas as it almost used up the last of the beer.

Thanks to the thoughtfulness of our good friends, we now had some ready cash and lots of nice household items that would come in handy as soon as we obtained a place of our own.

The following week we found work on a dairy farm north of Stonewall which would keep us busy for several months. I received $150 per month while Margo was paid $50 per month for cooking and housekeeping.

Near the end of September we drove north trying to decide what to do next. What really concerned us was the fact that she was pregnant already and we didn't even have a roof over our heads. Back at my parent's house we found Ruth, her husband Ken and their baby boy Bobby visiting from the coast. Over the course of the weekend we decided to head west to the coast with them. Three days later we were on the road on the first big adventure of our married life. Our friends didn't want to see us go and Kemmy Miller told me we'd be back in God's country by Christmas.

We left Winnipeg early on October 10 and reached Vancouver two days later in the midst of Typhoon Frieda. That storm knocked the electrical power off in the city for days.

The next day we drove to Powell River which was a new experience as we had to take two ferries to reach that isolated town. Ken and Ruth graciously allowed us to stay with them in their two-room rented house until I could find work and get a house of our own.

We were thankful that we had a roof over our heads until I received my first pay. However, the living arrangements were far from ideal. We were sleeping on a mattress on the floor in the tiny kitchen which meant that we couldn't go to bed until the others had retired to their bedroom for the night. The first weekend was kind of a nightmare for us when we realized Ken and Ruth ran with a wild and rough crowd, the likes of which we had never seen. These were guys that Ken had grown up with and he insisted on hanging out and drinking with them, even though he was out of his league with that type of dangerous men.

We sat up all night as the party wore on. As we lay on our mattress watching daylight creep in through the window, we talked about what we should do. We were so homesick for Manitoba that we would've left that morning if we'd had enough money to buy gas. We could've found a way but I kept remembering Kemmy's words about how we would be home before Christmas. I was determined not to give up until we'd given it our best shot.

Within five days, I'd found a job for a company called Mahoods. There I worked hard at the dangerous job of setting chokers on a high-lead logging operation. It would be easy to write a whole chapter about my experiences on that job. I must mention that I was earning the princely sum of $17.68 per day which was more than I could've imagined in my wildest dreams. We found a small house to rent just two short blocks from Ruth and Ken's. We moved on October 27 and for the first time we felt like we were really getting somewhere.

Our bubble burst when heavy snow in the mountains put an end to my logging job. By then I had formed a plan with one of the young men that I worked with. On November 17, he and I packed my car and took the small ferry to Texada Island to work at picking wild Blue Huck brush. The beautiful dark-green waxy leaves were in great demand to be used in floral arrangements and we could make as much as forty dollars a day. The bad part was that we had to live in old shacks that weren't fit for the mice that were crawling through them. We often fell asleep in damp clothes in makeshift beds on the floor. I ruined my caulked boots when I fell asleep one evening while trying to dry them in the oven of our little wood-burning stove. That was a long, sad night for me as I lay there in the dark with the wind howling and the cold rain slashing down. There was lots of time to think about my huge responsibility. I had a young wife and a baby on the way and I was feeling very lost and alone while longing for my carefree life back home. It didn't help to know that I had received a letter from a man in Lundar offering me work for the winter fishing season. Only my stubborn pride kept me in BC.

One day in early December, I left the car with my partner and took the short ferry ride back to Powell River to see my sweetie. To my surprise, I found she was in the hospital in severe pain. The doctors were quite sure she was going to lose our baby and the next morning she did miscarry. Margo was feeling like a failure and was grieving for the baby that we would never get to know. A kind doctor took the time to explain that this was often nature's way of dealing with a foetus that might be abnormal in some way. I tried to comfort Margo by telling her that we were young and there would be lots of other babies when the time was right. That wasn't what she needed to hear right then but it was the best I could do.

I knew my partner would be waiting for me at the ferry dock on the island that evening so there was no choice but to walk on that ferry and leave Margo with Ruth. I fell asleep in that filthy shack with mixed emotions. I felt so sorry for Margo but I was secretly relieved that I wouldn't have to be providing for a baby so soon.

January 7, 1963 found me back in Powell River trying to make a living by picking Salal brush for the same company. I couldn't work with my partner any longer even though I would be missing the good money we'd been bringing in. Also I missed Margo and couldn't stay away another day.

On January 23 I began working for the municipality on a winter-works project installing sewer lines in the Cranberry area. I found myself meeting some real good men and forming true friendships that would last a lifetime. The job was temporary so we saved every penny that came in. We ate pancakes day after day with the syrup cut in half with water to make it go further. Margo was proving to be a wonderful cook and the best part was that she never made me do the dishes. She insisted that I rest after my hard day's work. It was like living in Heaven with my best friend and lover.

As my winter job was winding down I obtained work on a salmon troller, thanks to Ken. A young man named Axel Ostrom had a brand new fifty-foot boat and he needed a man for the summer fishing season. Ken had accepted a position on another boat so he suggested to Axel that he give me a call. On May 4, 1963, my twenty-first birthday, I boarded the Karen O with my sleeping bag under one arm and my duffel bag slung over my shoulder. I was starting a new career that would last three years and provide enough adventures to fill a whole book.

A few days later we reached the port of Prince Rupert which was where we would deliver our fish to the Co-op fishery. One sunny morning, a rarity for Prince Rupert, I was walking up the steps leading to the dock when I met a young man on his way down with a box of groceries. He introduced himself as Konrad Swanson, originally from Bella Bella. He now lived in Vancouver and was paying for college by working summers for Collin McKea, owner of the salmon troller, Hike. Within minutes I knew that I really liked this young fellow. We promised to get together over the summer any time our paths happened to meet. I'd made another friend for life that morning and I would share many pleasant experiences with him over the coming years.

I truly loved the work of fishing aboard the Karen O but hated the forced separation from my wife. I did see her once that summer when the fishermen went on strike on July 14. Axel and I arranged to have his wife Audrey and Margo fly up to join us in Prince Rupert where the boat was tied during the week-long strike. Axel and Audrey stayed at his cousin's house while Margo and I had the boat to ourselves. By now she was pregnant again and expecting a baby in January of '64.

The season lasted until the end of August and a few days later Margo and I were on our way to Manitoba for a visit. We were on top of the world. Margo was coming along in her pregnancy and feeling well and we had acquired a good-sized bank account. We had enjoyed our year at the coast but didn't like feeling trapped by the ferries which were a necessary evil if you wanted to live in that town. I remember the wonderful feeling of freedom I experienced when we finally left the mountains behind as we headed east with nothing but wide-open spaces ahead of us.

That was a wonderful vacation spent mostly on our parents' farms. I helped Dad get the hay home and did lots of hunting as well. All too soon it was time to head back west. This time we had company, a young friend of ours, Sandy Miller. Not quite eighteen, he came along to see if he could find work.

The following year was almost a repeat of the previous one with a few major highlights. We had let our rented house go when we left town but we soon found a tiny house owned by a good friend, Les Ortloff, who lived next door. When Sandy found work, he moved over to room with Les.

One memory stands out in my mind about that little house. I was finding that my wife had a mischievous streak that caused her to play little tricks on me. She was always jumping out at me from behind a door or some dark corner and nearly causing me a heart attack. At first I laughed it off, but as time went on I started warning her severely that she was going to be sorry if she didn't stop this bad behaviour.

One night she went to bed while I sat up for a few minutes to finish reading the paper. When I put the paper down and walked into our bedroom, Margo sprang out from behind the door with a horrible scream. She scared me so badly I struck out at her with a side-handed slap to the top of her head. Her hair was done up in tight pin curls, as was her habit, and bobby pins scattered to the other end of the room. She stood there in disbelief for a few seconds before she asked, "What did you do that for?" I quickly replied, "I don't know. Why did you jump out at me when I've asked you a dozen times not to do that anymore?" We were very close to having our first major squabble but instead of fighting we suddenly broke into peals of laughter when we realized how ridiculous we both looked. That was the last time she ever scared me like that and it was the only time I ever struck her.

I'd built up a good work record the previous winter so had no trouble getting work for the municipality again, this time installing sewer lines in the Westview area. I had time off to have my appendix out but was back on the job within three weeks.

Margo's pregnancy went well and on the twenty-ninth of January 1964, our daughter Donna Lynn was born. It had been a hard labour and Margo was exhausted after a three-day struggle. We were delighted with our little girl and the rest of the winter passed quickly as we learned the parenting game. Margo was a natural mother and there was no more alcohol for her, though I still liked my beer whenever I could afford it.

When fishing season arrived, we decided Margo would spend the summer with her parents and mine in Manitoba. It would save us paying rent all summer and she wouldn't be so alone like she had the previous year. Early in April, I was on the Karen O heading north to the Fairweather grounds in the Gulf of Alaska. We had fished those fertile grounds the previous year and wanted to be there when the season opened on April 15.

We made it there for opening day and found only two American boats for company. The first day passed quietly dragging our trolling lines and lures through the water in hopes of finding a school of hungry salmon. It was getting late in the evening when Axel suddenly jumped out of the cock pit in the stern. He went running past me into the galley and disappeared through the hatch leading down into the engine room. A second later, he stuck his head up and yelled that we were sinking and that I was to get on the deck pump and start pumping for my life.

Axel had sensed the boat was riding low in the water and was becoming sluggish. When he went below, he found several feet of water sloshing around with the bottom part of the engine already covered.

I grabbed the handle, inserted it in the pump and began pumping as fast as possible. A second later, I heard the gas engine on the auxiliary pump start up and a steady stream of water began shooting out the side of the boat. With both pumps working, the water level began to drop until Axel was able to feel around in the icy water and locate the problem, a broken sea cock on one of the water lines. He put in a temporary plug that kept the water out until the boat was drained dry and then was able to change the broken fitting.

We were at least forty miles from the nearest land and well out of sight of the two other boats. If the darn thing had to break, at least it happened at a good time. An hour later we would've both been sound asleep and the episode might've had a far different ending.

That was a very long summer. The only contact I had with Margo was through one phone call and the letters and pictures we exchanged. We were committed to saving money so we could buy my parents farm some day. The strange thing is that I still drank heavily at every opportunity without giving a thought as to what it might cost me in years to come.

Around the end of August we fished our way down the coast until we reached Axel's home town of Sidney on Vancouver Island. In Powell River a few days later, I packed my car and drove to Manitoba to see my family. It was strange to see how my little girl had grown in five months. Margo and I were happy to be together again and so much in love. I believe she was pregnant within a week of my arrival.

Now that we had some money in the bank we decided to purchase a 1963 four-door Pontiac. It was a big brute and underpowered with a six-cylinder motor but it was roomy and comfortable which was what we needed for our growing family.

All too soon we were back in Powell River and the routine continued. We were now in a basement apartment that gave us more space. I found new work for the winter, driving machines for my friend, Jim Dyck. We had worked together for the municipality but he had branched out and bought an ancient construction outfit. Jim was only a little older than I and he was a real going concern, a non-stop talker and hard worker who was determined to make a go of it on his own.

That was a very exciting winter learning to operate heavy machinery but it also brought more close calls into my life. There were several times when I thanked God for sending a guardian angel to watch over me and keep me safe.

April, 1965 found me hugging a very pregnant Margo goodbye and heading back out on the boat for what would prove to be my final season on the water. I loved the fishing but hated being away from my family for such long periods of time.

The fishing turned out to be extremely poor, not at all like it had been the previous year. It seemed that we spent most of our time running from one area to another trying to find those elusive salmon. By the middle of June we were back on the Goose Island grounds, not that far from the northern tip of Vancouver Island. One day we heard there was a terrific storm coming which was going to last for several days. Axel decided to head for Vancouver instead of going back to Prince Rupert.

We reached Powell River early in the morning of June 18. Axel dropped me off at the wharf with instructions to meet him in Vancouver in a few days to unload the fish. I phoned home but was surprised when there was no answer. I called Ruth and she told me Margo had delivered a fine son just a few hours earlier. She loaned me her car and I drove to the hospital where I found my mum in the waiting room. I was sporting a good beard and it was fun to walk up to her and realize that she didn't recognize me.

Margo had an easy delivery which only lasted a few hours so she was on top of the world when she saw me walk in. Kevin Daniel was a strong lad with a set of lungs that would wake the dead. I spent the morning with Margo and the afternoon with Donna who was being cared for by neighbours, Tom and Louise Wilson. At first she was very shy but it wasn't long before she was hanging right on to me again.

That evening I went partying with Sandy Miller. He was living with Brian Kaartinen and several other lads from Manitoba. The party lasted until the early morning hours when Sandy suddenly disappeared. We found him sleeping peacefully at the bottom of a trench that had been dug out across the yard. That is how I celebrated the birthday of my first son—out drinking with friends all night. It was foolish and thoughtless, especially when I had precious little time to spend at home.

A few days later Axel and I were back on the water going through the motions as we put in the last two months of a dismal season. I wasn't unhappy though because even with the fishing as poor as it had been, I still made very good wages. When I finished fishing we headed east for a good, long holiday.

* * *

Late that fall my dad badly injured his knee by jumping off a tractor after raking hay in the cold wind for hours. A few days later he was in so much pain he was admitted to Eriksdale Hospital where he spent the next month. I was anxious to get back to the coast and get a job lined up for winter but couldn't go away and leave Dad with no help. His leg improved enough to allow him to move back home but he was confined to bed for several more weeks. It looked like his farming career was at an end and he became very depressed.

He was worried about what was going to happen to his beloved farm after he retired and he asked if I was ready to buy him out. We had saved a considerable amount of money but it was far from what we needed to get started farming. I told him there was no way I could do it so he spread the word that his farm was for sale. It wasn't long before the buzzards were circling and he had several offers to buy. When it looked like he was going to make a deal with one real estate agent, I suddenly had a change of mind and asked him if there was some way we could work out a deal. I couldn't abide the thought that someone else would own this farm that I loved so much.

A deal was struck that would allow Margo and I to rent the farm and lease his cows for two years to see how we made out. By then we would know whether there was a chance that we could make a decent living there.

On December 13, we left our children with Mum and went west for our belongings. It was a sad time for us as we made the rounds of good friends in Powell River to say, "Goodbye and be sure to come visit us on the farm."

We arrived back in Manitoba with all our worldly possessions packed into a little U-Haul trailer towed behind our car. We were home for good. A new career was about to start, one that would see lots of hard times before things got better. It would build lots of happy memories too, the kind of memories that would last a lifetime.

FOUR
Life on the Farm

January of 1966 found us settled in and living with my folks. Dad's knee was gradually improving but he was still depressed and far from well. Mum was in her glory with two grandchildren to cuddle and spoil to her heart's content. She and Margo got along very well.

After three years of steady income we knew things were going to be much different now, so we made up our minds to make some changes in our lifestyle. Cigarettes had risen to the unbelievable price of sixty-five cents per pack. Between the two of us we were going through three packs a day. There would be no money for such foolish indulgences so we quit smoking on the first day of the year.

For the past three years I'd been drinking booze at every chance I got. With no steady income to support my bad habit, I turned to brewing my own wine and beer in our bedroom above the kitchen. It seemed I always had a large crock of brew bubbling beside the warm chimney that ran up through a corner of that room. To this day, I don't know how Margo put up with me for the next few years.

One evening late that first winter I decided to visit my neighbours, Kemmy and Gert Miller. Margo's brother Art had been staying with us for a few days and he came with me. I guess guardian angels come in all shapes and sizes because he was my angel that night.

Kemmy and Gert were happy to have us back in Manitoba so they always had a warm welcome for us whenever we stopped in. I brought along a gallon jar of my strong, homemade wine and over the course of the evening I made a pig of myself, drinking way too much as usual. Art finally convinced me to go home and let those good folks go to bed. He did the driving and practically carried me into the house when we got home. Margo tucked me into bed where I passed out for an hour or so until I woke up sick. I was too drunk to get out of bed so I leaned over the side of the bed and threw up on the floor.

The next morning my wife was up early and cleaned up the mess before the kids had a chance to get into it. She should have dragged me out of bed and mopped the mess with my hair but she took care of it the only way she knew how. I wish I could say I learned my lesson right then and there but the truth is, my behaviour got much worse before it got better.

By early spring we were feeling the need to have a house of our own so a building bee was held to erect a small house not far from the big house. A few years earlier, Dad and I had cut a bunch of poplar saw logs and had Ralph Dodge come in with his portable sawmill to rip them into rough lumber. I still had my share of that lumber and it was used to frame the house. The neighbours pitched in and within a week we were living in our own two-room shack. One room was used as a kitchen and living room while the other was the bedroom for us and the two kids.

Margo was happy with her very own house but I wasn't feeling happy at all. I was feeling ever so sorry for myself and trapped by circumstances beyond my control. We were facing a money crisis where it might be years before the farm was producing enough to allow us to live in the manner to which we had become accustomed.

To make matters worse, I'd received an offer of a job on a halibut boat on the west coast and it was tempting. After each trip, every boat had to tie up for an eight day lay-off and that would allow me time to fly home and get some farm work done. I was seriously considering heading out there and when I mentioned this to my parents one day, my dad offered me some advice. This was the only time he ever did so, either before or after that day. He told me I was going to have to make up my mind whether I wanted to farm or fish because there was no way I was going to be able to fish and do a good job of farming. This was very sound advice and it put an end to the silly notions I'd been entertaining.

Still the whole situation looked to be totally impossible. The farm was only developed enough to support a herd of forty cows and the price for a large calf in the fall was only sixty dollars. My portion of the herd had expanded to seven cows and we leased Dad's cows at seven dollars per cow for each calf she produced. The land was rocky and rough and mostly covered in heavy bush, while the old machinery was in bad shape. My solution to these problems was to drink myself into a stupor as often as possible and try to forget about them.

By late spring Margo was becoming tired of my childish behaviour and there were times when we would argue bitterly. Sometimes, days would go by without a kind word being spoken between us. Soon all I could think of was getting away from the whole mess by going back to my carefree life as a fisherman where money and booze flowed like water and there were no problems like I was facing here.

One morning Margo and I got into an argument that left us furious with each other. I stormed out of the house and went to work, fuming about how unfairly life was treating me. Finally I made up my mind I couldn't stay any longer with a wife who wouldn't talk to me half the time. I went back to the house where I grabbed a suitcase and started packing. Margo followed me into the bedroom and asked me where I was going. I replied that I was leaving her. I didn't care where I went or what happened to this dead-end farm.

That was about as far as I got with the packing because as quick as I put something in the suitcase, Margo snatched it back out again, all the while pleading for me to stay. It still hurts me to think how frightened she must've been at that moment. It wasn't that she was losing any great prize but I guess I was all she had and she didn't want to lose me. We had two little kids to care for and there was no way she could do it on her own.

About a week later, I left home early one morning to work on the fence out at our lake pasture. Green grass was sprouting everywhere and it wouldn't be long before the cattle could be turned out to graze. In a swampy area near the lake shore, a mass of yellow blooms was growing in clusters in among the willow trees. The wild Marsh Marigolds were in bloom, with their beautiful, yellow blossoms in sharp contrast to their dark-green waxy leaves.

The thought crossed my mind that on the way home I should pick a beautiful bouquet for my wife. As quickly as the thought had come, another inner voice asked, "Why would you want to do that? You know she hates you so much she won't even talk to you most of the time." I was obviously still at the stage where I wanted to blame someone else for all my troubles, most of which were my own making.

That evening I finished my fencing project and was heading home when I noticed those beautiful blooms beckoning to me. Almost as if I was drawn there against my will, I got off the tractor and picked a huge bouquet which I carried home in one hand. Margo must've been watching for me because as soon as she saw me get off the tractor with the flowers, the house door flew open and she came running across the yard. I barely had time to mutter, "These are for you," when she hit me full force and hugged me so tightly the flowers were crushed against my chest. We kissed and we cried and that night we really talked for the first time in ages. The love was back in our lives again. It was a turning point in our marriage and things slowly improved after that. That simple bouquet of flowers became a ritual continued every year without fail.

By fall, Dad was well on the road to recovery. He'd bought a new pickup truck and installed his homemade camper on the back of it. Mum and Dad packed it with enough food and fishing gear to last all winter and headed south to Florida. Margo and I moved into their house where we had more room to move around.

Things were looking up a bit. We sold our calves for enough money to pay the bills we had run up in the grocery stores and at the machinery dealers. I applied to take part in an adult upgrading class held in Eriksdale. It was government sponsored and paid a small wage which would take care of our living expenses for the winter. If I applied myself to my studies I would be rewarded with my grade ten equivalent in the spring. My cousin Ted and I agreed to take turns driving the fourteen mile trip to town to cut down on costs. It also gave us a chance to get to know each other better as grown men.

I had no intention of doing anything except put in my time and collect my paycheque every two weeks. To my surprise, it wasn't long before I was engrossed in learning in a way that had never happened before. Each night I would pore over my homework in order to be ready for my next test. This book learning was going to be valuable when it came to keeping the farm account books in good order. It seemed wise to try to learn as much as possible for the duration of the course.

Christmas came and went and we received an occasional letter from Mum telling us how much they were enjoying this trip of a lifetime. They were basking in the sunshine and doing some fishing while traveling wherever the wind blew them. It seemed things were as good as they could get.

On February 8, 1967 I was in class one afternoon when I was called to the phone. It was unusual to be getting a call, so I went quickly. Margo was crying as she told me she'd just received a call from a doctor in Clewiston, Florida telling her my parents had been involved in a motor vehicle accident. Mum wasn't expected to survive for very long. Dad was seriously injured as well but his injuries weren't life-threatening. She told me to get home as quickly as possible in order to be there when the doctor phoned again.

Running out of the building into a cold, sunny afternoon, I jumped into my car and headed for home with tears filling my eyes. As I drove I prayed to God to save my dear mother who had so much to live for. I tried to bargain with God when I told Him if He would spare Mum's life, I would be forever faithful to Him. I promised I would start going to church every Sunday and would dedicate my life to Him. My pleading was for nothing because I got home in time to get another phone call from Florida. Mum had passed away in the hospital.

I went out of the house to do the evening chores and I really lost it. I ranted and raved, cursing God for taking my mum from me so early in life. I asked Him why, if He had to take someone that day, He couldn't have taken someone bad instead of Mum who had never done anything to hurt anyone in her whole life. She had dedicated her life to her family and friends; she was generous to a fault and would always give freely to someone needy. That night I was so angry at God; I turned my back on Him when I needed Him most.

Within a day or so, we had things more or less under control. Rose was living in Winnipeg and she had boarded a plane for Florida to bring Dad home. She also made arrangements to have Mum's body shipped to Manitoba so we could have a funeral. I was doing my best to make all the necessary arrangements. I phoned Reverend Jim Scott and asked him to hold the service for Mum. I was trying so hard to be a strong, brave man when I actually felt like a little boy who wanted his dear mum back.

I also had to arrange to dig the grave in the cemetery at Scotch Bay where Mum's parents were buried. It was common practice in those days for all the neighbours to get together to dig the grave by hand. A couple days after Mum died, I phoned my Uncle Will who was in charge of looking after the graveyard and asked him to meet me to measure out the grave site. He said I would need four wooden stakes to mark the corners of the plot, a crowbar to dig holes for the stakes in the frozen ground and a post hammer to drive the stakes in. When I hung up the phone I told Margo I was in trouble because my crowbar was lost and hadn't been seen for months. It was most likely lying flat under several feet of snow and wouldn't be located again until spring. Walking out of the house I headed for the work shop to have one more look for that wayward bar. The cold was vicious and the snow squeaked and crunched under foot. Halfway across the yard, I felt a gentle but firm hand on my right shoulder and heard Mum's voice as plainly as if she'd been standing beside me. The soft words she spoke made perfect sense to me as she said, "It's right over there, son." The gentle pressure from her hand turned me to the right and my eyes focused on the top of the crowbar where it stood leaning against the power pole that carried electricity from the yard pole to the house. Only a few inches of the bar were showing above the snow but thanks to Mum, I now had the bar I needed.

More than that, I'd been given a sign that Mum was okay and she had stayed around to keep an eye on her little boy who needed her so badly. I went slogging my way through the deep snow, gasping and crying all the way. When I reached out and touched that cold bar, I sank to my knees and cried bitter tears until I couldn't cry any more. That should've been the defining moment when I praised God and thanked Him for His mercy and grace. Instead, I stubbornly kept my heart closed to Him.

The next morning dawned very cold and clear at 41 below Fahrenheit as I left for the city to meet Dot, Kathy, Ruth and Margie who were arriving by train from the west coast. It was a long, sad trip home to the farm that day.

Rose and Dad arrived in Winnipeg the night before the funeral. My Uncle Art and Aunt Mary Watson brought them out the next day. I walked out into the yard to meet Dad and was shocked to see him looking like a frail, old man. I put my arms around him and hugged him with our cheeks touching as I told him how sorry I was. All he could say was, "It's a bad business, boy," before he limped off to the warmth of the house. He was far from well and as soon as the service was over, he was admitted to the Eriksdale hospital.

I wish I could say I experienced some miraculous revelation in church that day but the truth is I sat there sullen and silent, refusing even to say the Lord's Prayer. An hour later we stood out in the graveyard, surrounded by friends and family, and watched the casket disappear into the frozen earth. The minister said a prayer and then we all stood around hugging each other for a few minutes. I stayed behind because I'd helped dig that grave and wanted to help cover it, to be there for Mum right to the end.

As we prepared to start shoveling the dirt back into the hole, a man who had been helping the undertaker pulled a small flask of whiskey out of his pocket. Taking the cap off the bottle, he offered it around to, "Help warm your insides up, boys." How I wish I'd been brave enough to take that man by the scruff of the neck and toss him out of the cemetery but when the bottle was passed to me I took a tiny sip. This happened at the graveside of my dear mum who hated boozing with all her heart. To this day I am ashamed of myself for not standing up like a man when I had a chance to do so. It would've been one last fitting show of respect for the woman who had done everything in her power to teach me right from wrong.

Ruth stayed with us for a couple of weeks until Dad was well enough to travel and then she took him back to the coast with her. We were alone again and life gradually returned to normal. I threw myself into my studies and managed to do very well in my exams when the five month course was finished. Margo had been a real trooper and besides caring for the two kids, she had looked after the cows extremely well through calving season. It had been an effort in co-operation because I often found myself feeding hay in the dark after getting home from school.

That summer was a real test of our endurance when we had to take the kids with us to the hayfield day after day. We made a little camp and had Donna watch the baby while we worked nearby. Somehow we got through the summer with no major bumps or bruises and the haying got done in good time.

Every weekend it seemed we were partying or at least I was—Margo was being a good hostess. She enjoyed the company and always put on a nice lunch about midnight before she retired to our bedroom. With two little kids to look after she had to get her sleep while she could. I would drink with the gang until the party broke up as daylight arrived. I would fall into bed and sleep most of the day before waking with a miserable hangover that made me hard to get along with.

I became worried when my friends starting telling me about different things I'd done in the night. Often I would have no recollection of the incident until maybe a few days later when it would return to my mind like a flashback. I knew enough about alcoholism to know I was suffering from blackouts. This scared me so badly I made up my mind to limit myself to four drinks per night. At the next party I put my plan into action. When I reached my limit of four drinks I switched to coffee for the rest of the evening. It seemed almost comical, but sad as well, to see friends becoming tipsier as the night went on and laughing about silly things that weren't funny at all. I felt like a fish out of water and at one point I walked over to the table and prepared to pour myself another drink. Something held me back.

The party ended and I was able to help clean up the house before we went to bed. There was no hangover the next morning and from that day on I kept my promise and never had more than a few drinks. Married life continued to get better and we could slowly feel mutual trust building in us again.

The following year Dad married Mary Kirker, a nurse who worked in the Eriksdale hospital. They bought a trailer and lived in town for a year before moving to Salmon Arm, BC, where they built a small house on five acres. Dad was soon busy growing a large garden. They formed new friendships and enjoyed many happy years together.

Margo and I decided we loved farm life enough to try to make this land our permanent home. A deal was made with Dad where we would buy the farm by making payments every fall as soon as we sold our calves. Dad gave us a very good deal on the farm. In fact he handed the place to us for a pittance. We never missed a payment and when the last instalment was made, I wrote a letter to him offering to continue to make those yearly payments for as long as he should live. Dad refused to take another penny but he must've been deeply touched because he kept that letter of intent. It was found among his possessions many years later.

Those first years on the farm were extremely lean ones. Cattle prices were low and a good-sized calf was still bringing only sixty dollars in the fall. Our total income only amounted to a few thousand dollars a year. To help make ends meet and to feed my growing family, I took on any kind of work available. Early each spring, I fished with a trap net in a nearby creek for spawning suckers. These rough fish were sold to mink ranchers in Lundar and the surrounding area.

In the summer of 1970, we were finally able to take a trip to the west coast to visit family and friends. It had been close to five years since we'd left and it felt so good to reach the coast and see and smell the ocean again. Our friends and family were happy to see us and we received warm welcomes wherever we went. The time passed very quickly. All too soon we had to say goodbye and head for home again, stopping overnight at Dad and Mary's house in Salmon Arm.

The cattle prices were gradually improving and the herd had grown considerably, so in the spring of 1971, we took a huge gamble and sold enough bred heifers to pay for a new truck. It was a 1971 GMC with an eight-cylinder motor and the shiniest blue and white paint we'd ever seen. That brand new truck cost us a total of $3195 but it was an excellent truck and would last for years.

We also purchased a new side-delivery rake and a used chore tractor. With little Donna helping us, the haying went very quickly that year. It seems hard to believe those kids were driving tractors and raking hay at the tender age of seven years but they were and they did a darn good job too.

In the fall of 1971, I was able to obtain a commercial fishing license. This allowed me to fish with gill nets under the lake ice during the winter months. Early in the season I fished using only a small sleigh which I pulled by hand. When the ice became thick enough, I started using our truck. The first heavy snowfall put a halt to the operation. With nets in the water, I had to find a way to get back on the job. The solution came in the form of a brand new 1971 snowmobile. Purchased at a price of $700, it put a very large hole in our savings account but there was no other way. I had to get out there and lift my nets.

When the following winter arrived, I was back on the lake, hoping to make a living by doing an honest day's work. My hopes were in vain because the sought-after pickerel were very scarce that year. I soon realized I wasn't even making enough to cover my expenses. By year's end we realized we were expecting another child. With our family growing larger, I had to find a way to bring in some extra dollars.

There were coyotes in abundance that winter and the prices were still high, so I pulled out my nets and turned to hunting, determined to get all I could as soon as possible. With experience, both my hunting skills and my aim improved greatly. There was seldom a day that I came home empty handed. There would be at least one and sometimes as many as three dead coyotes tied to my machine when I arrived home at night.

With Margo expecting our third child we started dreaming about buying a better house and moving it to our farm. The old house was in bad shape and every time it rained, the roof leaked in numerous places. I didn't let that bother me too much as long as it didn't leak over my bed but poor, pregnant Margo would climb out of bed and place pots and pans under the drips to avoid having to clean up a mess in the morning. She was using cuss words that she usually kept in reserve for very special occasions.

Mid-June brought a terrible rain storm that could have had serious consequences for our son, Kevin. I would like to quote several paragraphs from a letter I wrote to my dad.

"June 15

Talk about extreme weather conditions! We are getting another downpour tonight. The kids and I were out west picking roots this afternoon and it was sweltering at about ninety degrees above Fahrenheit. Early this evening, a real dirty-looking thunderstorm came up from the west. When it started getting too close, we hit for home to help Margo with the chores.

We barely got the chickens under cover when it started to rain. Lightning was flashing and thunder rolling. Just as Margo, Donna and I were putting the geese in their shed; a bolt of lightning came down and hit the house before passing down the lightning rods into the ground. The noise is hard to describe—sort of a ripping, tearing noise followed by a tremendous bang. The shock of it almost knocked me down and I wondered if we had been hit.

Kevin was in the house standing by the sink and said when the bolt hit, the lights went out. He got a light shock and told us there was a blue light flickering all around the edge of the metal sink. He was a scared little boy and wouldn't go near anything metal after that.

I want to thank you, Dad, for having the foresight to put lightning rods on this house. If it hadn't been for that, Kevin would most likely have been killed."

That storm gave us a first hand lesson in how farmers are totally at the mercy of the weather. We are always too cold or too hot, too wet or too dry and at times like that there is precious little we can do except whine.

Cattle prices were much improved and for the past two years we had signed a contract with a large ranch. We bred all our cows artificially to the exotic Limousin breed which was being introduced to North America. We were guaranteed a price of $200 for bull calves and $250 for heifers which was almost like winning a lottery. Our herd had been expanded to about seventy cows so we had some real money in the bank for the first time.

Now that the kids were old enough to help me in the hayfield, Margo was able to take life a little easier that summer. As she approached her due date around the end of August, she was still going out twice a day to milk her old cow. She'd barely gained any weight during her pregnancy and we started to think the baby was going to be very small.

On August 29, Margo started showing strong nesting instincts by house cleaning and getting things in order. The next morning she was awakened by the first labour pains and I took her to the hospital in Eriksdale that afternoon. I came home to stay with the kids.

Late that evening I phoned the hospital and was told labour was progressing slowly. I went to bed but didn't get much sleep. A huge thunderstorm had swept in from the west, the rain was pouring down and there was water running all over the floor in several places. That reminded me of Margo and the way she worked so hard to keep our house neat. I made up my mind right then we would have to find a better house and move it to the farm.

If I thought I was having a restless night, it must've been much worse for Margo as she endured endless hours of misery without even a doctor to help her. There had been a car accident and Doc Paulson was busy all night caring for the young lads who had been badly injured. Hulda Larson was called in to take care of Margo through that long night. Finally the doc was able to find time for Margo and with his assistance, Steven Robert came into this world. He weighed a whopping ten pounds, one ounce so no wonder he had taken his time coming.

The night we gained a son, our good friends, Mac and Linnea Miller, lost their son, Wade, in the car accident. It was hard to feel happy about my new son when I saw how very sad Mac was feeling. A few days later I helped a bunch of friends as we dug a grave for Wade.

Things weren't going well with our new baby because he had developed pneumonia. At the age of three days, he had to be rushed to the Children's Hospital in the city. Our neighbour, Sandy Miller, drove us in his car with nurse, Hulda Larson, in the front seat holding the baby. Margo and I rode in the back seat holding hands. It was hard to believe our big, strong son could be so sick, so early in life. A few hours later he was placed in an incubator in the preemie section of the ward. He looked out of place beside those tiny premature babies but this treatment saved his life. He recovered quickly and was soon home with his family.

Margo became very run-down from trying to do too much and she developed an infection. I took over the night care of the baby and as a result, a strong bond developed between us in a way that hadn't happened with the two older kids. I loved those kids too but I was very rough on them. It shames me to think about all the times they got a spanking for doing nothing more than acting like any other normal kids. They were quite a pair, those two, and their mischievous nature got them into all sorts of trouble with their mother until she would order me to step in and straighten them out. I would let things slide until my temper boiled over and then would give them severe punishments that almost equalled the one bad beating my dad had given me.

Margo's health improved just in time for her to be my nurse when I came down with a very bad case of mumps late that fall. Margo was left with the care of her new baby and all the chores with only the two bigger kids to help her. She had driven tractors plenty of times but she had never run the loader which was used to bring hay to the cattle. Kevin was only eight years old but he'd spent hours riding with me on that tractor. He was able to teach her how to use the various levers. One day she left one of the levers in the wrong position after she had dumped the last load of hay. When she speeded up the engine to come back to the yard, a hydraulic line blew out and hot oil blasted her in the side of the head. She came to the house dripping with oil and when I laughed at her, she was almost ready to kill me.

Things got much worse when the mumps left my face and migrated to other more tender areas of my body. I lay in a feverish trance all weekend until Margo got Gert Miller and her son, Craig, to load me up and take me to town to see the doctor. Dr. Baffsky admitted me to the hospital where I was cared for until my fever broke several days later. All the while, my dear wife kept things together on the farm.

Now that I had pretty well quit the boozing, it seemed like we didn't get much company anymore. We certainly didn't have time or money to spend on a social life. My painful episode in the hospital caused me to do some serious thinking about how fragile life really can be. There were times when I believed if I died the next day, I didn't have enough friends to make it worthwhile having a funeral. I was also having second thoughts about things like God and Heaven and whether I would ever see my departed loved ones again.

In Salmon Arm, my dad was having the same kind of thoughts. My step-mother Mary had a very strong spiritual faith and she made sure Dad attended church with her on a regular basis. As he got older, his letters home were filled with worried thoughts as he struggled to believe in the same way Mary did. He wanted to believe in God but he was full of doubts and didn't know how to reach out to ask for help. That was exactly the way I felt, at least part of the time, especially when I was feeling weak and vulnerable. When I was feeling strong and healthy, I figured I was doing quite all right on my own.

The spring of 1974 arrived late and wet with constant rain which was the last thing we needed right then. We'd spent the winter shopping for a house we could move in but anything we could afford wasn't a lot better than the house we were living in. That problem was solved when my cousin Ben offered to help us build a new house. With the help of relatives and neighbours working together under Ben's supervision, the house went up in a week and that included a full basement. I was working myself to exhaustion day after day but my friends were doing the same thing with no thought of remuneration.

Once the house was closed in with windows and doors in place, we went back to doing our regular farm work until after haying was finished. By mid-August we were working at putting on dry wall. Once again we had good neighbours helping out, sometimes until after midnight. Dot came home and she and Margo did a major part of the plastering while I worked at hauling hay home to get ready for the coming winter.

On November 4, we moved across the yard into the new house and it was like a dream come true, especially for Margo. She was pregnant again as I'd convinced her that the episode with the mumps in my nether regions had made me unable to father any more children. Our youngest daughter, Tracy Leanne, proved me wrong when she arrived on July 22, 1975. Margo was feeling fine and was totally enjoying raising our brood in our new house.

* * *

Sandy Miller and I became partners during the winter fishing season and we purchased an old, narrow-gauge bombardier. This large snow machine on rubber tracks and skis allowed us to travel back and forth from home to the fishing grounds in total comfort. The fishing was quite good and we worked long, hard days.

Every spring for a number of years, I hired myself and my old diesel tractor out for spring seeding on one of the big grain farms in the area. This farm was owned by Nathan Finkel. Sixteen hour days, at ten dollars an hour put far more money in my pocket than I was making on my own small farm. Nate supplied the seeder and all the fuel for my tractor so it was a winning situation for me. Some days it was very hard to stay awake on the tractor

With a decent income at last, we decided to spend at least one thousand dollars every year to clear the bush off the land in order to grow more crops to feed our growing herd. A thousand dollars would clear a lot of land when the going rate for machine and operator was only twelve dollars per hour.

Once we had new land cleared, we often found ourselves working from dawn to dusk preparing that land for seeding. Margo worked beside me hour after hour picking roots and stones and piling them on the flat rack of a rubber-tired trailer. She was incredibly strong for her size and never seemed to tire, no matter how hard the work. The older kids were big enough to work along with us and it was surprising how much work they could do if you made a game of it.

Every time we became discouraged we could always look back at what we'd accomplished already. It was fun to dream big and imagine what this farm would be like some day in the not too distant future. I realize now that Margo had it so much harder than I did. She put in long days in the fields and then went home to do all the barnyard chores before going in to make supper. I stayed in the fields until dark before coming in to eat and fall into bed, exhausted. Margo would stay up until the dishes and the last load of laundry were done.

In July 1976, we were shocked to hear that my Aunt Mary and Uncle Jack Ross had been killed in a terrible car accident while coming home late one evening from a fishing trip. Also killed were their two young grandsons and their neighbour, David Lawrence, who had been riding with them.

Several weeks later, I was outside my shop one Sunday morning doing some welding. I was thinking about the Ross family and how you could be here one minute and gone the next. These sad thoughts set up a fierce longing in me to go visit our family members on the west coast, most of whom we hadn't seen for six years. Dropping my tools, I walked to the house and asked my good wife, "How would you like to drive out to the coast?" "Sure," she answered, "When would you like to go?" I was all excited and told her I'd like to leave the next day.

By Monday evening, everything was ready. We'd borrowed a camper which we mounted on our truck. Everything was packed for an early start the next morning but we were too excited to sleep so we left that night and drove halfway across Saskatchewan before stopping for a few hours' sleep.

We reached my Dad's place in Salmon Arm the next day and spent a few days visiting with him and Mary. By the weekend we were on the coast, where most of my sisters lived. The following weekend, the majority of our family had gathered at Rose's home in Prince George. After several days of visiting, we left late one evening to drive as far as McBride, BC, where we pulled into a campground for the night. The next morning after breakfast on the tailgate of the truck, we headed east again.

A long day's driving took us through Alberta and almost across Saskatchewan. After supper, five of us jammed into that tiny camper for one last sleep on the road. By 6 AM I was wide awake and raring to head for home. Margo and the kids were still sleeping soundly, so I quietly sneaked out to avoid waking them. In the cab, I woke Donna and asked her to scoot over a bit to make room for me. She was soon wide awake and we talked as we rode along, both of us excited about getting home. By eight o'clock, we were nearing the town of Roblin, Manitoba, traveling on a narrow paved highway. A glance back through the window into the camper told me Margo and the three younger kids were still sleeping. Not being especially hungry, I decided to keep driving until they woke up.

In a field on the south side of the highway, several flocks of geese were landing. The hunter in me became aroused and I kept sneaking glances at them until something on the highway ahead caught my eye. An oncoming car was starting to veer across the center line. I sat up straight and gripped the steering wheel in both hands as the car continued to move further into my lane on a path that would take it across our lane and over the shoulder into the ditch unless it hit us first. I made a split-second decision to do something that went against everything I'd learned about safe driving. It was too late to head for the ditch on the right hand side. At the last second, I hauled the steering wheel hard to the left, to what should've been the other driver's side of the highway. There was a howl of screaming tires as the truck laid over almost on its side. The two vehicles missed each other with only a few feet to spare. I caught a glimpse of an old, white-haired man as we flashed by. He was staring straight at me with terrified eyes. There was no time to think about it as I made a sharp turn to the right to get straightened out again. The passengers in the camper were rolled to one side of the bed and then the other.

With the truck heading straight down the road again, I sneaked a quick look in the mirror. The car was skidding along the shoulder of the road with gravel and grass flying in a great cloud of dust. He was a good driver alright and managed to bring his car under control and back onto the pavement. He pulled over to the proper side of the road where he stopped on the shoulder. I did the same thing before looking back through the window into the camper. It was a wild looking bunch that stared back at me. With her eyes wide as saucers, Margo was yelling something at me and it definitely wasn't ladylike.

I walked around to the rear to open the camper door. We enjoyed a group hug while I tried to explain what happened. The car was still sitting down the road and Margo wanted me to go have it out with the guy. I told her, "He's an old man, honey. We'll just let him go on his way." I didn't trust myself to go back and speak to him in a rational manner. I was afraid I'd lose my temper and give him a thrashing. Looking at my family whose lives he had jeopardized made me want to weep. We continued on our way, leaving that car still there on the shoulder of the highway.

Many times I have wished I'd gone back to talk to him, to find out what caused him to do such a stupid thing. Had he been driving all night and fallen asleep at the wheel? That was quite possible. Was he dying a slow cruel death from an illness and decided to put an end to his misery by committing suicide? That wasn't very likely. Was he a hunter like me and spent a few seconds too long admiring the flocks of geese landing in the field beside the highway? That was a real possibility. Maybe he was having a heart attack which caused him to lose control of his car. I may never know the answers to these questions.

* * *

The next few years passed in a blur of activity. Sometimes it seemed as though there were only two things in the world for us, work and hockey. Kevin's team was becoming very successful but we had to travel many miles to make that happen. Some teams in our regular league were as far away as eighty miles and we often traveled hundreds of miles to take part in tournaments. As manager of the team I was required to go along to all of these tournaments. Margo continued to support me in these endeavours even though there were times when she was left alone to care for the farm for several days at a time. Those hockey seasons were the worst because I was trying to juggle many things. There were the regular farm chores to look after and I was still fishing full time on the lake. Worn out at the end of a long day's work, I would drag myself to town twice a week to take part in the boys' hockey practice. It was all worth it when I watched my son play hockey. He was the smallest lad on the team but he was becoming a very good skater and a proficient scorer. Standing in the bench with the team and helping with the line changes, I would hoot and holler and encourage the lads as much as possible. It felt so good to see them win. Sometimes, however, we lost sight of what it was all about. We never seemed to be satisfied with a loss, even if the boys played their hearts out against a much tougher team.

After each exciting game it would take me hours to settle down and sometimes I wouldn't sleep at all until the following night. Even then my sleep would be fitful, with wild dreams of hockey flashing through my mind. It never occurred to me that I was courting disaster with this seemingly innocent lifestyle.

1979 was very hectic with three kids in school. Both boys were playing hockey and with Donna in figure skating, it was difficult to meet all the schedules. At times like that we co-operated with neighbours to share the driving. That summer passed quickly in a blur of work. It was hard not to feel envious as I sat on my tractor in the fields and watched the endless stream of cars passing by on their way to the beach. It seemed we were always too busy or too tired to go down for a refreshing swim in the evenings like we'd done earlier in our marriage. The beach brought more commitments because all the kids were taking swimming lessons which meant juggling precious time between the lake and the hayfield.

In early spring of 1980, I started experiencing severe pain in my chest and left shoulder. These attacks of pain always seemed to happen after I'd watched an exciting hockey game. At first I laughed it off when Margo questioned me. I told her it was over-excitement. Finally the attacks got too severe to endure, and I reluctantly consented to go to the hospital for a check-up. Dr. Baffsky took it seriously and immediately ordered an EKG which showed my heart was as healthy as a horse. He prescribed mild tranquilizer pills to help me through the stressful periods and they worked like a charm. As soon as the hockey season was over and we settled into the summer routine, there was no more need for pills.

That fall Tracy started attending Kindergarten every second day. That brought quite a change in our lives. Margo was free to become my fishing partner in the winter fishing season. I bought out Sandy's share of the bombardier and Margo came with me every second day while Tracy was in school.

December 8, 1980 was a sad day for us and our neighbours when we lost Kemmy Miller, barely sixty years old, to a fatal heart attack. It felt like I'd lost a father, a friend and a big brother all in that horrible moment. Kemmy and Gert had done so much to help us and without their help we wouldn't have been able to stick it out on the farm during those first lean years. He'd always been there for us when we needed him and I wondered how we were going to get along without him. There was precious little sleep for us that night as we lay there and talked about the good friend we'd lost much too soon.

After Margo dozed off, I lay awake wondering if Kemmy had believed in God and if he'd made it to Heaven. My own health scare the previous spring had been causing me to do some serious thinking about how fragile life is and what would happen to me when I died. I didn't believe in God or Heaven but I was starting to wonder if there really was such a place as Hell.

The fishing that winter was extremely good and we were putting in long days. We were making more money than we ever dreamed possible. We earned enough to buy a new car, the first new vehicle we'd bought in years. Margo loved fishing, especially when those fat pickerel were popping out of the hole one after another. After all those years of always being around our kids, we finally had time to spend totally with each other. We talked for hours as we faced each other across the net and we were becoming closer than we'd been in any part of our marriage. I believe this was the first time I truly made up my mind to totally commit myself to making our marriage as good as it could possibly be.

That winter was terribly hectic, what with hockey, farm chores, calving and fishing to attend to. I felt like my nerves were going to snap when I lay awake at night hoping for merciful sleep to give me rest.

One evening in the spring of 1981 I was stricken with a severe chest pain that traveled down my left arm. It was so bad we were sure I must be having a heart attack. It eased a bit after a few minutes but I was unable to eat.

The next morning I had another severe attack and this time we headed right in to town to see the young doctor in charge. I was given an EKG which didn't show any abnormality but the doctor was convinced I was suffering from angina pain. He gave me a nitro-glycerine pill under my tongue which didn't do anything for the pain in my shoulder though it did give me a severe headache. I went home convinced I was on death's doorstep.

Over the next few weeks my condition worsened while the doctor tried me on everything from vitamins to anti-depressants. Nothing worked and the attacks continued daily until I was unable to do a decent day's work. On May 9, we received news that my brother-in-law Norm Josephson had died from a severe heart attack. We knew his health wasn't good but this latest news came as a shock. I was sure I'd be the next one to go and the thought of leaving Margo alone with four young children was unbearable.

I'd been scheduled for a stress test in the city but the nurses were on strike. It was early June by the time a new test was scheduled. I was a total wreck. I was taking prescription tranquilizers and soon became worried by the thought of becoming hooked on them. The minute I went to bed my mind would kick into high gear. My thoughts would race from one worry to another and sleep would be impossible without the help of the pills.

Margo managed to look after the kids as well as the farm chores. I lost close to forty pounds and weighed less than 180 pounds which made me look like a walking skeleton. My state of mind was so bad I didn't want to see or talk to anyone except Margo and the kids.

I reached a point where my whole body was suffering from such severe cramps that the muscles in my back were standing out like baseballs. The pain was so intense, it was becoming unbearable. The nerves in the back of my legs were moving like there was a bunch of worms crawling around under my skin. On the day of my stress test Margo drove us to the city because I couldn't bear the thought of driving in heavy traffic. The test showed my heart was in very good shape. The doctor who ran the test asked me if I was a professional athlete because my heart rate returned to normal so quickly.

That night Margo and I talked long and late, trying to make sense of what happened to me. By now we knew the problem was all in my head and we feared it wouldn't be long before I'd end up in a mental hospital like my brother had. I made a vow to Margo that night that if she would stick with me, I would devote my life to making her happy. Promises like that are a lot easier to make than they are to keep but time would prove I had made a major step toward maturity that terrible summer.

Haying time arrived and I managed to find the strength to climb onto the tractor every day to put in long hours that had become an ingrained habit over the years. Every day, I found myself feeling stronger. By the middle of August, I was only taking one pill before bedtime just to sleep. It took another month before I was off the pills entirely and several months before I really started to feel normal again. My appetite returned with a vengeance and my weight soon improved. Winter came early that year and when the fishing season opened on November 1, Margo and I were on the frozen surface of the lake setting nets long before daylight. The fishing that season was fabulous and just what I needed to gain full recovery.

The following year passed with no major incidents or special excitement. The farm was doing well. The cattle herd had expanded to eighty cows, which was all the farm could support. I was still enjoying hunting every fall and gradually formed a strong bond with a good friend who became my hunting partner. Lorne Smith was a partner with his father Gordon and his brother Ray in their lumber yard business in Eriksdale. Lorne loved hunting as much as I did and we seemed to compliment each other's style. He was an extremely careful hunter and a good sport as well as being a crack shot. He had three sons and Lorne and I enjoyed the experience of passing our knowledge and hunting skills down to our kids.

* * *

The summer of 1983 brought the fulfillment of a life-long dream when I was able to make arrangements to obtain a private pilot's license. It was early August before we got started with lessons. I was able to arrange for a one hour lesson in the early morning and another hour just before dark. That allowed me to go home during the day to get haying done. In the evenings we attended ground school in a classroom in Lundar School. It was a very exciting time as I went about my work with my head filled with mental notes from the morning lesson.

Within a few days I improved my landing skills and progressed to other exciting procedures like stalls and spins. The stalls were easy but the spins were something else. The plane would be deliberately put into a spin and I would find myself staring straight out the windshield at the ground spinning in circles several thousand feet below. It was unnerving at first but I soon grew to love it and easily mastered the steps needed to recover from the spin.

By mid-October I had my license. A month later, we purchased our first plane and that began an eighteen year love affair with flying. Len Heroux was good enough to come check me out on my first few flights before I was left to get some solo experience. Steven was the first family member to go flying with me while I practiced touch and go landings on skis in our snow-covered hayfield west of the barn. I was extremely cautious while learning the ins and outs of flying but there were still times when my inexperience almost caused an accident.

Early winter brought a close call when my snow machine broke through the ice on the lake and dove to the bottom. I was soaked through but managed to swim out onto the ice. A nearby fisherman drove over and gave me a ride to shore.

Margo had her turn at testing the temperature of those frigid waters a few days later when she walked onto an area where the ice was very thin and fell through with a huge splash. Luckily the water only came up to her armpits. Steven ran over and grabbing her by the hood of her parka, dragged her out. It looked exactly like an Eskimo hunter dragging a seal out of the water and that image was so funny I made the mistake of laughing at her. She headed straight for shore and jumped in the truck to go home to change her clothes. I called to her to make sure she came back before dark to pick us up and the language she used on me was certainly amazing. To give her credit, she was back there waiting for us when we got off the lake that evening.

The following summer, 1984, my Uncle George Cowdery died peacefully in his sleep in his house near Eriksdale. He left his farm to us as well as his small herd of cattle and the machinery. The land consisted of five quarter sections and was a welcome addition to our farm, turning it into a small ranch. It didn't come easily though, as we would spend many years building fences, clearing bush and turning the rough land into good pasture. Margo worked countless hours by my side through all kinds of weather, as did our two youngest kids.

January 12, 1985 was a sad day for us when we lost a good friend, Peter Kaartinen, to lung cancer. He hadn't been well for quite awhile and he lasted only a few weeks after being diagnosed. He left an empty spot in our lives and we would always remember the wonderful times we spent with him and his wife, Harriet. He'd been a very good fiddle player and I'd spent many hours making music with him over the years.

The next few years we continued to improve the farm and build up the cattle herd. Steven and Tracy had taken over as helpers when Donna and

Kevin left home and we had some decent machinery. My life was almost stress-free. We'd even started going to an occasional dance and were meeting new friends.

Ruth had been visiting us on a regular basis from her residence in Powell River, BC. She was seriously considering moving back to Manitoba because she'd never felt completely at home on the west coast. Her marriage had broken up and her two kids, Bob and Sheila, were young adults on their own. We told her if she wanted to move back, we would build a basement for Uncle George's house and she would have a home for as long as she wanted.

In October 1987, Margo and I flew to the coast to help Ruth pack her belongings into a large U-Haul truck. I drove the large truck and Ruth followed behind driving her half-ton pickup with the camper jammed to the top with all manner of stuff she couldn't bear to part with.

On the way home we stopped to visit Dad for a few hours in Salmon Arm. He'd lost Mary several months earlier and was terribly lonely. He wanted us to stay overnight but as usually happened when I was heading home I had the bit in my teeth and wanted to get as far down that highway as possible before it got dark. How I wish now we'd taken the time to spend the night because that was the last time I saw him. Perhaps he had a premonition because after he played his fiddle for us, he had a serious talk with me and told me he would will his beloved fiddle to me if I promised I would learn to play it. I told him I couldn't promise I would learn but did promise him I would give it a very good try. He smiled at me and it reminded me of the special smile he gave me when I was a little boy.

The weather turned very cold and we arrived home to be greeted with several inches of heavy wet snow. Winter had arrived early and it was a struggle to get the truck close enough to unload Ruth's belongings. Several neighbours helped us unload the truck and that made light work of it.

Life settled into a winter routine but it didn't take long to realize that by giving Ruth a home, we had been blessed with a real God-send in return. She found work in town but was usually available to help us on our farm especially when problems arose during calving.

The fall of 1989 brought sadness to our family when Dad died. He had been ill for quite some time so it wasn't unexpected. Rose had moved Dad up to Prince George so he could be near her. According to his wishes, he was cremated and his ashes brought back to Manitoba to be buried beside Mum. He also wanted half the ashes scattered across our farm. It wasn't possible for all of us to get together that fall so we made arrangements to hold a service the following summer.

On July 22, 1990 we held a graveside service for Dad as we laid him to rest beside Mum in our cemetery at Scotch Bay. I'd avoided going there since Mum left us but now they were together I felt easier about visiting the graveside. It was comforting to imagine they were together again and would remain together for eternity.

The following day I was joined by Dot, Rose and Ruth as we drove around Dad's old farm, scattering ashes in his favourite spots. Kathy and Margie had been unable to make the trip back for the service. We took turns choosing spots and saying a little prayer for him. We spent several pleasant hours enjoying being in each other's company and telling stories about Dad. I was the last to choose and I led our group to the old ditch leading from our Black Pond across a high ridge to the lake. As a boy I'd spent many happy hours with Dad as we worked at deepening the ditch every spring. This ditch had been dug by my grandfather Watson and his sons and it was one of my favourite places. I knew Dad always loved to see and hear running water so I was sure he loved this beautiful, shady place as much as I did.

The sun was shining brightly with a few fluffy, white clouds scattered across a perfect blue and gold sky. I shook the last of the ashes out of the bag and as a gentle breeze carried them away I spoke to Dad saying I hoped he was happy now and approved of this special spot I had picked. The words were barely out of my mouth when a deafening clap of thunder came out of that blue sky, so loud it frightened us. The thunder rolled across the sky and receded into the distance as we stood gazing dumbly at each other. When the last rumble died away, I found my voice, "Wow! I think he must've approved." I felt shaken by the incident and I'm sure my sisters felt the same. Somehow it made me feel better, as if Dad had tried to show us he was safe and happy wherever it was he'd gone.

The winter of 1990 brought a wonderful opportunity when we were invited by John and Pat Fjeldsted to travel with them by car to Florida for three weeks. Margo and I had never had a vacation together without kids tagging along so we jumped at the chance. I had rather mixed feelings about it because Mum had been killed in Florida in that accident but we agreed this was too good an opportunity to pass up.

We left early on Boxing Day and drove slowly south, taking almost six days to reach North Port, Florida. I kept a daily travel log which I sent home as letters to Steven and Tracy. They were left in charge of the farm with Ruth keeping an eye on everything in general, including the kids.

We moved into a condo belonging to one of John's friends and were soon outside enjoying a cool drink with the thermometer reading 80 degrees Fahrenheit. It wasn't hard to see why my parents loved Florida and it made me feel good to know Mum had been happy during the last few weeks of her life.

On January 6, I placed a call home the way I promised I would do every Sunday. Ruth answered and told me some devastating news. The previous afternoon Sandy Miller had been riding a snowmobile when it rolled over and crushed him, injuring him fatally. We were struck dumb by this horrible news and I considered hopping on a plane and heading home to see if I could be of some help to Sandy's wife, Audrey, and their five sons. We finally decided to carry on with our plans and head to the Keys to visit friends Ray and Betty Larson.

We reached our destination shortly after noon and Ray introduced us to Betty, his lovely wife, a former nurse as well as a cancer survivor. She is the kind of person who makes you feel you've known her all her life. Ray was very happy to see us but his happiness was dimmed when we told him the news about Sandy. Ray had often hunted on Sandy's land and had known him well.

That night I lay awake thinking about Sandy, wondering if he was safe in Heaven or if he'd even believed in God. As well as we'd known each other, we'd never discussed that topic. Once more it had been driven home to me how precious and fragile our life is. I tried to say a prayer but felt like such a hypocrite, I gave that up as a bad job. I also questioned God, asking why, if He was so merciful, He allowed this to happen. If I was expecting an answer, I was out of luck because the night dragged on with no miraculous happening. Finally I got up and poured my feelings out in my daily journal. That letter still sits in my desk, stained by bitter tears shed for my friend that night.

* * *

Back home, I began another hobby that soon grew to be a favourite with both Margo and me. I inherited Dad's fiddle as promised and was in need of someone to help me learn to play. Margo had been attending a weight loss class in Lundar and had become good friends with a wonderful lady, Mabel McLeod. Her husband Fred was a good fiddle player so I got up my nerve and asked him to help me get started. That was the beginning of another friendship and it wasn't long before I was spending Monday evenings playing music with Fred while the women were at their meeting.

In the spring of 1993, Tracy graduated from high school and that September she moved to the city. The last chick had left the nest and we couldn't have been more delighted. We'd been in the child-rearing business for close to thirty years and now we had the whole place to ourselves.

FIVE
On Our Own At Last

When Tracy left home, our friends must've found us a bit odd when they tried to sympathize. Our response was that we totally loved our new freedom. That was the key word all right. We had freedom to come and go, and freedom to do what we pleased, whenever we chose to do so.

We'd become very fond of old-time dancing, often traveling miles almost every weekend to attend dances at our favourite halls. Those dances were just like the dances we'd attended in our younger years but without the fighting and drinking. We met many wonderful people and more close friendships were formed.

We were still working as hard as ever. Our cattle herd continued to grow but we had Ruth helping us and the three of us got along great. At times we did more laughing than working but we always managed to get the job done. We called ourselves the Three Stooges because of the silly antics we pulled on each other, making all that hard work seem like fun.

I continued playing the fiddle and both Margo and Ruth were learning to play the guitar. We spent hours in our kitchen practicing until we had a few tunes that were passable. At first neither of the women was able to follow those fiddle tunes because it seemed impossible for them to know when they had to switch from one chord to the next. Most of the tunes were played in the chords of D, G and A. We soon worked out a system where I would use head signals to let them know when to change. A tip of my head to the left meant that they were to change from D to A. A tip to the right brought them back to the original chord. Tipping my face upwards instructed them to change to G. Just like when we worked together, we seemed to do more laughing than we did playing music.

Calving was very labour intensive with the season starting around the end of January. Extremely cold weather meant the cows had to be checked at least every two hours, night and day and I was running myself ragged running back and forth. We solved that problem by building a small, insulated bedroom inside the barn. Flood lights overlooking the maternity pen helped me keep watch in between snatching a few winks of precious sleep. I was always able to fall asleep at a moment's notice and that came in handy during calving time.

We also installed an intercom system from the barn to the house, making it easy to call for assistance. I was able to handle routine work on my own but sometimes needed help with a difficult birth. I would call to the house and my wife would sleepily answer. No matter how tired she was, it never took her more than a few minutes to appear at the barn, sometimes with her nightgown tucked into her jeans.

In August 1994, Ruth bought ten head of bred heifers from us, thus becoming a proud owner of her own small herd. We supplied the feed and care for her cows while she supplied labour on the farm. It turned out to be a system that worked extremely well for all of us.

* * *

In November 1994, a dream came true for Margo and me when we took a two week trip to Hawaii. Traveling with us were my sisters, Dot and Kathy, and Margo's sister, Bertha, and her husband Dave Allen. We met them in Vancouver and we soon found ourselves high above the ocean heading west on an adventure of a lifetime. I found myself rather in awe at the size of this immense body of water. About five hours later, the island of Maui appeared off to our left. As we neared the island of Oahu, we saw Waikiki beach and the small extinct volcano called Diamond Head. After several years of looking at pictures of this beautiful place, I was actually seeing it with my own eyes. Happily our vacation went just as we'd dreamed. We swam and had picnics and enjoyed the beautiful weather on Kauai and Oahu.

That winter we had new excitement in our lives because we were expecting our first grandchild. Donna and her hubby, Brad Brassor, were living in the town of Fort Saskatchewan, just north of Edmonton, Alberta. They became a family of three when a fine baby boy, Daniel Bernhard Bradley, was born on April 26, 1995. He was named Daniel after me and I felt quite honoured.

A few weeks later we drove west to see our new baby. We went on from there all the way to the Queen Charlotte Islands to visit Margo's brother, Ken, and his wife, Lorna. We enjoyed several days' salmon fishing with Ken on his small charter boat. Ken is an excellent fisherman who works out of the town of Masset, BC, and we experienced fishing like you can only dream about. One evening while returning from the fishing grounds, we came upon the salmon troller Karen O lying on anchor. Thirty years later, there she was, looking good as new. The bow poles were gone but other than that it was like looking back in time. We pulled up beside her and spoke to the owner. I would've loved to have gone aboard but it was almost dark and we had a few more miles to go to reach town.

That night I lay awake thinking about what my life might've been like if I'd continued fishing instead of coming back to the farm. It wasn't hard to imagine that my drinking may have escalated out of control with all the easy-come easy-go money. Those first years on the farm had been really hard but they forced me to pull up my socks and tighten my belt and that most likely made a better man of me.

Back on the farm, things were getting better all the time. At times I felt like we were living in a bubble that was about to burst. It was almost as though the better things became, the more that nagging, little worry kept gnawing away at me, trying to tell me that this wonderful life was too good to last.

Early 1996 brought more happiness and excitement into our lives when we found out Steven and his wife, Linda, were expecting a baby around the end of August. They had bought a house in Eriksdale and we were looking forward to spending lots of time with that new baby.

July 20 that year brought a new chapter into my life when I started writing daily diary letters to my sisters on the west coast. Dot had been encouraging me for years to write a bit every day instead of infrequent letters sent on rare occasions. As each sister read the letter she initialed it and sent it on to another sister. After everyone had seen it, it was mailed back to me and placed in a binder to become a journal for my grandkids to read some day. The group grew to include my daughter, Donna and my niece, Sheila, and it was great to get letters from them in return. It helped all of us feel a bit closer to each other.

I started these letters just in time to record the death of my dear cousin, Edmund Watson, who died from cancer after a long illness. He'd been such an inspiration to me and was always ready and willing to lend a helping hand.

On the last day of August we got a call from Steven—he'd taken Linda to the Women's Pavilion in the city where her labour was progressing nicely. A little boy was born a couple hours after midnight. He missed being born on his father's birthday by only a few hours. The baby was named Kevin Byron after his Uncle Kevin and Granddad Lindell.

The following afternoon we drove to the city to see the new boy and we were amazed at how relaxed and comfortable Linda was as she handled that tiny baby. She'd always been a mother hen to her younger brothers and sisters so one more baby was a piece of cake.

Late September brought the beginning of goose hunting season but it brought worry as well, because my buddy, Lorne Smith, wasn't feeling well. He'd been suffering from a bad backache that never seemed to go away, and it was really causing him grief. He was so stiff and sore he could barely walk around while out on those early morning goose hunts.

Mid-January 1997 brought the terrible news we'd been dreading. Lorne had been diagnosed with cancer on one kidney and his lungs. The cancer was so advanced there was no possibility of treatment, either by chemo, radiation or surgery. Lorne was a fighter though, and he wasn't going to give up easily. I would like to quote from one of my hand-written letters that expressed my deep emotions during that awful period of time. (Edited passages of my journal letters, enclosed in quotation marks, will help shape my story from this point.)

"January 17, 8:10 AM

Lorne Smith found out yesterday that he has cancer in his kidney and both lungs. The doctor basically told him to get his affairs in order and be ready to die. The only possible ray of hope is an experimental treatment where they inject some drug into the affected areas. This supposedly causes the body's immune system to kick into high gear. They have to do more testing to see if he is a suitable candidate for this treatment.

My mind is just in turmoil and I can't quite believe this is really happening. I hardly slept all night, but when I did, I would quickly awaken and think that it was all just a bad dream.

Why do these terrible things usually seem to happen to such nice people? The good ones seem to be dropping like flies around me, and the rotten scum bags that go around hurting people seem to live forever.

The list of friends of mine that I've lost, that died young, just keeps growing. The Millers, Kemmy (heart attack) Mac, Wade, Ken and Sandy (accidental) are all gone. Our wonderful friend, Peter Kaartinen (lung cancer) and his brother, Dave (heart attack) are sadly missed as well. A friend, Jim Broadfoot from Woodlands, died two years ago at less than forty years of age, from brain cancer. Jim's father-in-law Ragner Steinthorson also died from cancer. I've lost two brothers-in-law, Norm Josephson (heart attack) and Ken Baxter, who died from MS. Then last year we lost our cousin Ed after a long struggle with stomach cancer.

These are only some of my friends listed, but the list goes on and on. What is the answer? The way we live must have a lot to do with it. Lots of smoking and various preservatives and chemicals in everything we eat and drink. Even the very air we breathe isn't pure any more. It makes one wonder what it will be like fifty years from now, if it's this bad now."

I was becoming uneasy when I contemplated the fact that we are all so vulnerable. More and more I found myself thinking about God and how I was going to have to stand and face Him some day. I was also keeping a wary eye over my shoulder to see if the grim reaper was closing in.

The stress I was under sent me to Ashern Hospital with severe chest pains on January 29. The doctor ordered an EKG and had several blood tests done. I was admitted to the hospital and hooked up to a heart monitor overnight to await the results. The tests came back negative and I was released early the next day. Within a few days we were in the thick of calving, which was the best medicine I could've had. By the time calving season was over, I had a firm grip on my emotions, deciding to try not to worry about things that were beyond my control.

As spring arrived we could see Lorne fading away. It was a very bad time for all of us. Even the wonderful news that we were expecting three more grandchildren couldn't erase the sadness we felt every time we saw Lorne.

We'd seen other friends dying from cancer and we never knew what to say to them. In fact we'd sometimes avoided these people for that very reason. We were determined not to let this happen to our friendship with Lorne. As the summer dragged on and his condition worsened, we often visited him and his wife, Donna, or they came to visit us. Lorne was using morphine to kill the pain and we'd get him settled in an easy chair where he'd doze off for awhile. Every so often he'd wake up and smile at us and maybe talk a bit.

I wanted so much to ask him what his beliefs were regarding God and whether he believed there was a Heaven. One evening, I finally got up my nerve and asked, "Lorne, do you believe in God?" He looked at me for a minute before he replied, "Yes, Dan. I do have faith." Then he closed his eyes and dozed off again. The opportunity for a close talk never arose again, as much as I would've liked it to happen.

On July 24, 1997, Kevin and his wife, Natalie, became parents to a son, who they named Steven, after his Uncle Steven. Now, we had two Kevins, two Stevens, and two Daniels in our family and it was getting a bit confusing.

On August 12, Lorne lost his battle with that cursed cancer, just three days before Donna and Brad welcomed a daughter into their family. Our emotions ranged from joy to deep sadness and then back again. That tiny girl soon sported a name that was longer than she was. Rebecca Ann Margaret Elizabeth Brassor was born a few weeks prematurely with her lungs undeveloped and for a few days her life hung in the balance. Named for her two grandmothers, Becky soon gained strength and became a picture of health.

When goose season rolled around, I was planning to ignore it but my friend, Fred McLeod, talked me into going out on opening morning. I knew hunting would never be the same without Lorne but life goes on and sometimes you just have to make the best of it.

We had our blinds set up and the decoys out in my barley field well before daylight. As I watched the light grow stronger in the east, I had the strangest feeling that Lorne was there beside me and I wondered if he approved of our set-up. We must've done something right because we had our limits of fat geese before the sun appeared over the horizon. Perhaps Lorne lent a helping hand and sent those geese winging straight to us.

We soon had another granddaughter, Cassandra Dawn, born on October 9 to Steven and Linda. For years we'd feared we were never going to get any grandchildren and now it seemed they were coming in bunches. We were really in our glory because we'd been chosen to look after little Kevin while Linda was in labour. The day after the birth, the three of us headed to the city to see Kevin's beautiful baby sister.

February 1998 brought the usual rush of new calves and after a month or so we were almost on our last legs. No matter how tired we were, life went on as usual and we even found time for a bit of socializing.

Our neighbours, Garry and Corrine Seidler were busy with their cows calving too, but they were also going through a terrible time as their fifteen-year-old son Tenason was fighting for his life against bone cancer. This young lad showed courage far beyond his years as he endured various treatments but he lost the battle on June 2, 1998. This was way beyond our understanding. It was bad enough to see friends our own age die, but this young man hadn't even had a chance to experience life. More and more often I found myself asking "Why?" but there was no easy answer available.

We welcomed another little girl into our growing family on November 23, 1998 when our daughter-in-law, Natalie, gave birth to Kristen Margaret. Now all three families had a matched set, with the boys being born first in each family. The kids were all healthy and it seemed as though life couldn't get any better. Again that nagging worry appeared in the back of my mind and I was thinking more about God and how I'd turned my back on Him.

I was using an old computer for my daily journal letters and was enjoying the new experience. I'd begun writing short chapters about various events in my life which I hoped would form a short book that would be nice to pass on to my kids and grandchildren some day. My Dad had done the same thing late in his life and we'd all enjoyed it so much.

Mid-December found me hard at work on the keyboard, tapping out a story about the journey I'd taken as a seventeen-year-old when Jim and I hitch hiked to Vancouver. It wasn't long before I realized there was way too much material for just one chapter. The idea was born to make it a book, complete with some old, treasured pictures. I believe the main reason I wanted to write that book was to let my children and grandkids know about their Uncle Bob.

Christmas and New Year's came and went and I barely noticed their passing. The hours flew by as I sat hunched over the keyboard typing in my awkward two-fingered style. February arrived bringing the first calves and the book took second place when I moved out to the barn bedroom. The book was foremost in my mind and I often found myself in the house working late at night trying to get a few more pages done.

The middle of February, 1999 found me close to finishing part three and late one evening I reached the point where I was telling about the last time I saw my brother. We were standing together on a bridge leading out of Kamloops, BC, as Jim and I prepared to head back to Manitoba. As I recalled how Bob and I shook hands and hugged before we said goodbye, bitter tears that I'd been holding back for years suddenly came pouring out. I wept as if there was no way to stop. It felt as if I was looking right into Bob's beautiful, blue eyes as we said goodbye. I realized in that moment that he'd already known he was going to take his own life. With that realization came the release of all the guilt and shame that I'd been holding back all those years when I blamed myself needlessly for his death. It took hours to finish the last page in that chapter but it was probably the most important hours I'd spent in my life.

When I finally dragged myself from the keyboard and walked out through the frosty night under the light of a million stars, I tried to pray to God that Bob was safe in Heaven. I wasn't sure whether people who committed suicide were allowed to enter Heaven but suddenly that seemed very important to me. Even more important was the realization that I wanted to be able to go to Heaven when I died. That made me start thinking about what I must do in order to earn God's favour and grace.

The following week I had an experience that changed my life forever and I wrote to my sisters about it.

"February 19, 1999

Things have been happening here that are beyond my understanding. Last week while writing about how Bob and I said "Good-bye" for the last time, I broke down and cried for a very long time. After it was all over, I felt calm and more at peace with myself than I've been for years. I now have a firm belief that somehow I will see Bob again some day. Maybe we'll be roasting wieners over the hot coals of Hell or maybe drifting around in fleecy, white clouds. I really feel that it will be the latter.

For the next few days, I mulled it all over in my mind, trying to make sense of the whole thing. I really felt I had found faith at last, but I wasn't too sure just what I had faith in. Finally I decided to just relax and watch for signposts along the way.

One day I asked God to give me a sign if this was all true and then just left it in His hands. That very night when I left my bedroom in the barn and went out to check the cows at 1:00 AM, I noticed that the northern lights were extremely bright across the northern horizon. It was a very cold, clear night with millions of brilliant stars but no moon shining.

At four in the morning I stepped outside for another pen check and stopped short when I saw a truly amazing display of northern lights, unlike any I'd ever seen. I ran back inside the barn to shut off all the flood lights before running outside again.

The sight before me was almost beyond belief. The lights had expanded until they encompassed the whole horizon in all directions and extended upwards toward the heavens. It felt as if I was inside a giant cone, like standing in the very center of an extinct volcano looking out through the hole in the top. The lights near the horizon were bright and clear, while, halfway up to the heavens they were a rich, red color that faded away higher up to such a brilliant white that I couldn't even see the stars shining through. At the very top, there was an opening with a golden circle of light framing half a dozen bright, twinkling stars.

I fell to my knees in the straw and prayed, "Oh, Lord, I asked you for a sign and here it is. How could I have ever doubted your love for me and how can I not believe that you are real?"

When I was finally able to pull my gaze from the sky, it had become so bright out I could easily see trees along the lake shore a full mile away. I felt extreme calmness and joy. The thought crossed my mind that the stars inside the golden circle were the faces of my beloved parents and brother and some special friends who had gone on ahead of me.

The lights stayed like that, as if they were frozen in a time frame for perhaps fifteen minutes before they started pulsating and flickering across the sky as they normally do. A few minutes later they were dancing across the northern horizon again.

My readers can make what they want of this but I took it for the sign I'd requested. I certainly never expected to see such a powerful message or have it arrive so soon after my request. I'll be watching carefully for more signs in the future."

For the next few days I walked around in a daze. I knew there must be a scientific explanation for what happened in the sky that night but I really didn't want to know. I'd asked God for a sign and within hours He had answered me.

It would've been wonderful if I'd become a true believer in God in that moment. The truth is that I continued on my merry way, believing that I was a good man and that should be enough. Time passed and the memory of that special sign in the night sky faded from my mind as surely as those lights had faded from my view.

The next two years slipped by in a blur of work with the occasional weekend dance to break the monotony. For some time we'd been talking about selling our plane because I never seemed to find time to fly. If the weather was good enough to fly it was good enough for work and work always came first. I gradually lost interest in flying and was only putting in ten or twelve hours a year. This wasn't enough to remain in good practice and I was trying to convince myself to sell the plane because it was just a money pit. In the fall of 2001, I made up my mind. Paul Wanke, a young man from West Hawk Lake was interested in the plane and we made arrangements to do the deal as soon as possible.

A few days later I took my last solo flight in my plane. I spent several hours cruising around this wonderful Interlake area that I love so much, waving my wings at various neighbours who happened to be outside looking up that day. Two miles south of our farm I swooped down to buzz the cottage of our old friend Linnea Miller. After being a widow for many years she was re-married to Adam Kell and they lived in her little house near the lake.

There was no sign of life there that day but I did notice in the woodlot beside her house, a bunch of huge ash trees that had blown down in a storm a couple of years earlier. At the first opportunity, I asked for and received permission to cut as much of that good wood as I needed. Linnea was only too happy to get it removed because she felt the dead trees were a fire hazard.

That winter Margo and I spent many happy days in that woodlot. We left home early each morning with one of us driving the dump truck and the other one following behind in the tractor. I used the chain saw to limb the trees and cut them into logs while Margo used the tractor to pull the trees up to the landing. She was nervous at first but overcame her fears and was able to use the loader to load all the logs into the truck. By noon we were on our way home with another three-cord load of prime firewood. Margo wasn't feeling very perky that winter and by the time we got home, she would be so stiff and tired she could hardly crawl out of the tractor. Of course I teased her about it, telling her that since she was several years older than me, she must be getting too darn old to cut the mustard.

That was a wonderful winter, one that I will never forget. Linnea would often show up driving her snowmobile with Adam riding behind her, hanging onto their little dog. They often invited us to the house for a bowl of soup. I always tried to repay them for their kindness by taking a bunch of smaller logs in the loader bucket to be dumped off in front of Adam's shop.

Twice we had campfire cookouts in the woodlot with some close friends on a Sunday afternoon. Linnea's brother, Albert Berg, had married our friend, Hulda Larson, and they were among the good friends gathered there. Linnea and Adam always showed up on their machine even though Linnea wasn't well. Her health was deteriorating daily and we suspected cancer though she wasn't saying anything about it.

Margo and I talked about how our friends seemed to be dropping like flies. She confessed to me that on her sixtieth birthday she experienced a premonition that she wouldn't live to receive her first old-age pension cheque at age sixty-five. As I was turning sixty on May 4, 2002, we decided to hold a dance in the Eriksdale hall. We invited our friends and relatives to help us celebrate my birthday and our fortieth wedding anniversary. The party featured a live band entertaining about three hundred guests. As we stood at the door and greeted friends and family, it occurred to me that not too many years earlier I'd wondered if we had enough friends to make it worthwhile having a funeral when I died. Now as I looked around me at all those smiling faces, my heart was filled with joy.

My five sisters presented me with a certificate that would pay for scuba diving lessons, another lifelong dream of mine. Within a week, I was enrolled in a diving class in the city and five weeks later I'd completed the training, passed my tests and had my certificate. It was another hair-brained idea but I had the full support of my loving wife.

That spring brought bad news about Linnea. Her health had grown much worse and she was confined to the hospital after being diagnosed with lung cancer. Margo and I went to see her one evening and we had to wait outside her room while the nurse combed her hair and got her looking presentable for visitors. She might've been down but she wasn't out and she still had her pride. The cancer won in the end and Linnea died on June 20, 2002.

Linnea's daughter, Darlene asked if I would do the eulogy at the funeral and I quickly replied that I would be honoured. I was fairly well acquainted with public speaking and felt sure I wouldn't have any trouble even though this was going to be a new experience for me.

On the day of the service we were at the Lutheran church in Lundar with forty-five minutes to spare. The church was almost full already but we managed to find seats near the back. As the minutes passed, I started feeling extremely nervous and was wondering if I'd taken on more than I could handle. By the time the service started I was a wreck and told Margo I didn't think I was going to be able to speak even if I found the nerve to go to the front. When the minister called me forward, I was shaking so badly my legs barely carried me.

I turned and faced the crowd and tried to speak and to my amazement the words wouldn't come out. I stood there for a couple of minutes staring dumbly at the sad faces until I managed to find my voice. I got through everything I wanted to say and then made my way back to my seat. I was so warm the heat was radiating off my body. I had no idea what had happened to me but I know I felt God's presence in church that day in a way I'd never experienced before.

No matter how hard we worked we always took time to go dancing and our circle of friends continued to grow but in the midst of our happiness came concern and worry. Ted and Linda's son, Darryl, was suffering from a large lump on the side of his neck. Barely twenty-seven years old and a hard-working single father of a little boy, Nicholas, he was going through extensive tests to try to diagnose the problem. When the results came back he was told he had a very aggressive type of fast-growing cancerous tumour and plans were made to get him started on chemotherapy immediately.

November 6, 2002 brought Margo's sixty-third birthday and a trip to Ashern to have a routine mammogram. Unlike a lot of women, this test had never bothered Margo as she took it in stride, considering it a necessary evil. When she got home that evening, I asked her if the test had hurt. She replied, "No, it never. Well, maybe just a bit on one side." That was all that was said until the next day when she told me that she now had a hard lump on her chest wall on the edge of her left breast. The lump was there all right, with hard, almost sharp, edges. I felt my own cold lump in the pit of my stomach as I looked at my dear wife gazing back at me with fear in her eyes. After talking it over we came to the conclusion that the machine must've been adjusted too close to her chest and had pinched part of the muscle.

The next day we were up early and on our way to British Columbia. The results of the mammogram wouldn't be back for ten days so we figured we might as well go ahead with plans we'd made. We had a good time on the coast but the worry and fear never really left us as we examined that hard lump before every bed time. We tried to tell ourselves it was nothing and kept our concerns to ourselves to avoid worrying our loved ones.

In Powell River we visited old friends and family members and my nephew, Bob and I enjoyed several dives in the ocean. This was a wonderful new experience for me. One morning we went out in Bob's boat to dive on an old wreck several miles off shore. Margo came along, as did Kevin and Nat and the grandkids. We all had a good time except for Margo who got terribly seasick from the motion of the boat. When Bob and I came up from our first dive, Margo wanted to get back to shore and we didn't take very long to get her there.

While we were on the coast we received word that another good friend, Benny Gleich, had died from cancer. Only one year older than I, he'd been very sick all summer but had somehow carried on with work as usual as long as he possibly could. He left behind his wife, Connie, and two sons, David and Jim.

Back in Manitoba we received wonderful news—the mammogram had come back clear. We chided ourselves for worrying needlessly all through our vacation but a week later, Margo showed me another small lump just under the skin between her breasts. We were extremely concerned and we visited Dr. Elzette Steyn in her Urgent Care Clinic in Eriksdale the next day. After examining Margo, she immediately booked an appointment with a specialist at the Manitoba Clinic in Winnipeg. Unfortunately, the earliest opening was not until January 7, 2003.

We settled down to wait through the holiday season with mixed emotions. We'd known for a short while that we were due to be grandparents again. Tracy and her husband, Rick Forsyth, were expecting a baby in June, 2003. Steven and Linda and kids came out for Christmas dinner and we tried to put our worries behind us and make it a happy time for all. At times I found myself laughing though all the while that nagging fear never left my heart. Every time Margo and I looked at each other, I knew the fear was eating at her too.

Dan and Margo in happy times.

SIX
The Devil Cancer

The first day of 2003 found us preparing to see the specialist at the Manitoba Clinic. There had been a cancellation and we were bumped ahead five days to January 2. We made plans to visit our good friends, Einar Jonnason and his mate, Elsie Wersak, on our way home from the city. We'd met this fine couple through our old-time dancing and had become very close friends over the years. When Elsie heard we would be stopping in, she insisted we should stay for supper. Knowing what a wonderful cook she was, we didn't put up much of an argument. We also planned on visiting Darryl Watson who was now undergoing cancer treatment in the Health Science Center.

For almost two months we had been hiding our fears from our friends and family until we knew exactly what we were up against. Perhaps the best way to convey the horror of that trip to the city is to quote directly from a diary letter.

"Dear Sheila, January 3/03

You have the dubious distinction of receiving the first letter of the New Year and I'm afraid it is going to be a real downer. Yesterday turned out to be the worst day in my life when we were told by a specialist that Margo has breast cancer. Margo said it was like getting hit on the head with a hammer and we are both still in shock. I know I have to be strong and brave for Margo's sake but all I really want to do is to break down and cry like a baby. That hasn't happened yet but I'm sure it will happen sometime soon.

Yesterday your mum came out to help me with chores as she has been doing often since Margo hurt her ankle about a month ago. Margo and I left for the city after lunch. We got there half an hour early and went straight up to the office in the Manitoba Clinic. The girl took Margo through to one of the examining rooms where she undressed and waited for half an hour for the doctor to see her.

I was almost asleep in a waiting room chair when I heard a man's voice call my name. It was the doctor beckoning me to join him across the room. The way he looked into my eyes made me realize it was indeed the bad news we'd been expecting. My legs got so weak; I could hardly walk. He took me in to where Margo was lying on the table and gave it to me straight out. My heart was beating so hard, the blood was pounding in my ears and I had to struggle to make sense of the words I was hearing. I caught snatches like, "I have to tell you that your wife has breast cancer... quite an advanced stage... adhered to the chest wall... not operable at this time... have to zap it with chemotherapy and radiation to shrink it... possibly operate to remove it later."

I felt like I'd been kicked in the guts and Margo told me later that my face turned snow white. The doctor asked me to leave the room so he could remove the small lump for a biopsy. I looked into Margo's beautiful, brown eyes and saw the terror that was in my heart reflected there. I stumbled back out to my seat in the waiting room with my mind in turmoil.

Half an hour later, Margo joined me and we hurried down to the lab in the basement to get a blood test done before the place closed for the day. A few minutes later, we went up to the main floor to have a chest x-ray taken. The doctor said we'll be hearing from him as soon as the biopsy results are in.

We hadn't been fooling ourselves over the past few weeks and I was expecting the worst case scenario that I could imagine, a cancerous lump in her breast and a complete mastectomy, then a full recovery. Now, who knows what to expect?

Seeing these words on the screen has opened the flood gates and I'm unable to do any more right now. Your mum just got here and Ray and Georgina Larson are on their way to see us. I really don't feel like I can face anyone right now but I'd better go in and talk to Ruth and Margo. Maybe that will help settle me down before the others arrive.

Later, same day, 6:45 PM

We had plans to visit Darryl Watson in the Health Science Centre but by the time we escaped from the clinic, we just wanted to get out of the city. I shouldn't have been driving but there wasn't much choice because Margo was more shaken up than I was.

Out on the highway, I phoned Elsie to tell her we wouldn't be there for supper. She got all excited and told me she had everything ready. She was just waiting to hear from us before she put the steaks on. I had to tell her the bad news and the poor gal was so shocked, she blurted out, "You must be kidding." I assured her that I wouldn't be kidding about a thing like that. She urged me to come out to their place even if we only stayed long enough to eat. We decided it might be better to get off the road until we had time to settle down.

Einar met us at the door and looked like he was in shock. When Elsie saw Margo she burst out sobbing and hugged us like she would never let go. We found we were able to eat a fair bit of the wonderful supper Elsie had made. She is so kind and generous, she packed all that was left and sent it home with us.

We were home in plenty of time to phone all our kids and Margie. We asked Margie to call my other sisters because it was hard talking right about then. Our kids were all shocked but I think the girls took the news better than the boys did. That confirms what I've suspected all along. Women really are the stronger sex, at least mentally, if not physically. Kevin was really shook up, as was Steven. Margo talked to Kevin and he told her that she is tough, that she had to be with the things her kids had put her through earlier on.

We went to bed and I actually fell asleep right away. About two o'clock this morning I woke up feeling like I was swimming up out of a nightmare. When I realized it was real, my heart started racing. I reached across to touch Margo and found that she'd gone to the other bedroom as she does sometimes when she can't sleep. Laying there for an hour with my mind whirling, I tried to figure out why things like this happen to the very best people but of course they just do. I did a lot of thinking as I lay there alone and I prayed to God to spare my dear wife. If I could trade places with her, I would do so this minute with no regrets. She really is one in a million and long ago I came to realize how fortunate I am to have had her for my mate these forty plus years.

I was barely sixteen when I met her and I think I loved her from that very night. If not love, maybe I was attracted by her wild and crazy, fun-loving ways. She really was different than any girl I'd ever met. Now that we've been together for so long, we are more like one person than two. If one of us is hurt, the other one feels the pain as well. I couldn't ask for a better companion for life. She is so kind and forgiving. I guess she had to be to put up with all the crap I put her through in the first few years of our marriage.

She has also been incredibly supportive. In the early years, she stayed home and looked after all the chores while I ran around the country to hockey tournaments with Kevin. Sometimes I would be gone three days at a time even during calving season. Thank God we had good neighbours who she could call on when she was in a bind.

When the farm finally started paying off, I began thinking about fulfilling life-time dreams. When I told her I wanted to become a pilot and buy a plane, she encouraged me to go for it even though it meant she had to do without a lot of things she wanted for the house.

When I got the bright idea of running sleigh dogs, she was all for it and guess who had to feed and water the wild pack all summer while I was in the hayfield. It was Margo just doing it without complaining as she always does.

I've done all the night shifts in calving season but I always knew I could call Margo any time during the night to help with a difficult birth. Sometimes she sounded half-dead over the intercom but she would always be out there in a few minutes, knowing quite well that she would likely get shouted at during the heat of the battle.

Every fall, I turn into a bit of a wild animal myself during hunting season but she has always encouraged me in my efforts because she knows how happy it makes me.

I've been buying a new tractor and loader every two years for the past twelve years or so. Margo's car is ten years old.

Last spring, it was scuba diving for me which with her fear of water must've been horrifying for her. All she said was, "Be careful and have fun."

If there is any bright side to this bad news, it is that the doctor mentioned he wants to use chemo and radiation to shrink the tumour and then he might be able to operate to remove it. At least that gives us some hope.

When my friend Lorne Smith learned he had a tumour on his kidney, there was absolutely nothing that could be done for him. We lost him less than a year later. We are praying they will find Margo's cancer hasn't spread and they will be able to shrink the tumour.

Jan 4, 8:20 AM

We managed to get through yesterday and things are settling down again. I quit crying before our guests arrived. Hulda and Albert arrived a few minutes before Larsons did and after group hugs, the women went into the living room while Albert and Ray joined me at the kitchen table. It must've been quite a sight to see three big, strong men sitting there with tears running down our cheeks as I told them what the doctor had said. It was hard to have company right then but at the same time it was so very good for us. We talked until eleven-thirty and then I had to get outside doing chores.

I phoned Ted later. He said Darryl was supposed to start chemotherapy yesterday. Darryl's little son has never had chicken pox so he isn't allowed to go into the isolation ward where Darryl is staying. That is so upsetting for both of them. Ted says Darryl amazes him with his up-beat attitude, saying that he just wants to get on with the treatments.

Our bad news is getting around and we've had an out-pouring of support from people who love us, both family and friends. In the past we often found ourselves avoiding friends who were in similar situations because we didn't know what to say to them. We know now that was a mistake. It's nice to know people are thinking about you when you need them.

Margo phoned up to The Pas and talked to Jackie Isfeld to see how she is recovering in her battle with breast cancer and to tell her about our troubles. The last we heard from Jackie was that she'd undergone surgery to remove part of her breast and was having lots of trouble with bleeding and infection. Now the doctors have discovered more cancer and they are going to take the whole breast off. If Margo's cancer was operable, we would've urged the doctors to do a complete mastectomy as quickly as possible. We have often discussed this very thing and I'd sooner have a wife with no breasts than no wife at all.

Margo got a call from our grandson, Dan, who asked her, "Gramma, do you have cancer?" Pretty hard stuff to take but it must be so hard for his little mind to get around this news as well. Dan is a very bright boy and he told Margo that it's lucky she didn't get it in her leg like Terry Fox.

Jan 5, 7:45 AM

We continue to get lots of calls of support and it really helps both of us to talk about it. I keep a brave face in front of Margo but find myself bawling my eyes out when I'm walking around doing my chores. I keep thinking about Mum and Dad working all their lives and then losing Mum to that accident just as they started their retirement. Ever since we started talking about retiring, we've felt like we're living on borrowed time. I guess, really, we are all doing just that. All we can do is stay in touch with our loved ones and tell them often that we love them.

Your mum gave up her half day at the post office to come out and be with us. She also brought Margo a huge bouquet of flowers from the shop in town. She is so thoughtful and we are truly blessed to have her for support. She and Margo get along like two peas in a pod.

A couple days ago we got a homemade Christmas card and letter from Heather Smith Thomas, the lady rancher down in Salmon, Idaho. The card showed a picture of their upper meadows at sundown. They live in a rugged but beautiful part of the country. We read her diary column in the monthly Grainews so had heard most of the news in her letter. Yesterday I wrote Heather a letter bringing her up to date on what is happening around here.

Margo is being incredibly brave and strong. On the surface, she is her usual cheerful self while I'm the one who seems to be having trouble dealing with all of this. I think I could deal with it much better if it was happening to me and not to her.

Jan 6, 7:25 AM

I phoned Margo's brother, Art, yesterday morning and he said he'd drop out later in the day. He showed up in the afternoon and had tears in his eyes all the time he was here. Before he left, he walked out with me to the maternity pen where I showed him the deadbolt for the gate and the guard for the water fountain that I built with metal he had given me. Before he left, we had a hug and then he stumbled off in the wrong direction. I had to call him back and get him started off toward his truck. He lost his very good woman, Dorothy, to cancer a number of years ago and I know he is experiencing some very strong emotions right now.

After supper, we got a call from Carolyn Miller saying that she and Craig would like to come over and watch the World Junior Hockey game with us. Canada was playing Russia for the gold and it had the makings of a good one. Their visit was just what we needed and we truly enjoyed having them."

Once word got around it seemed like the phone never stopped ringing, with good friends all wanting to lend support. That was just as well because I was putting in very long days fishing with Tiny Lamoureux. The perch fishing was excellent that winter and we were driving ourselves hard through all kinds of weather. The hard work helped keep me relaxed and functioning well but there was seldom a moment when I wasn't thinking about Margo and what lay ahead for us.

The afternoon of January 16 found us back in the city for Margo's appointment with Dr. Haiart. That had been a long two weeks. We were very anxious to hear the results and plan a course of treatments.

"We were there early and they took her in right away. The doctor soon came out and called me in so he could talk to both of us at once. He's a kind man but doesn't mince his words. He says Margo has had this cancer growing for a long time. He told us the blood test came back normal and the x-ray showed that her lungs and chest bone were clear. He said that was very good news. The test from the liver showed some abnormality but he thinks it might've come from the cholesterol pills she's been taking for several years.

His main concern was that the cancer may have spread to other parts of her body but it is unlikely that her lymph nodes under her arms are involved with this type of cancer. He planned to refer her to a cancer specialist at the Cancer Clinic in Seven Oaks Hospital and the specialist would order a CT scan. He said we have to keep bugging the doctors, including him, or we might have our records tossed on a shelf and forgotten.

He took out the four stitches from the biopsy which had confirmed that the little lump was in fact satellite cancer cells. I suddenly remembered to tell him about her ankle which has been sore for a couple of months. After he examined it, he sat for a minute with his chin in his hand deep in thought and then made a phone call to the Cancer Clinic. He explained the situation to the lady there and said he wanted an urgent priority put on this case.

He had told us we would be lucky to get in within a month but believe it or not, we are due back in there at ten tomorrow morning. The girl who runs the CT scan machine had one opening. We'll be feeding cattle at quarter to six in the dark but that is a small concern. I have been praying to my guardian angel to help speed up the process and he really came through big time. Believe in the power of prayer. I have been a quiet believer for quite some time but now my faith is growing stronger by the day. When I heard the words "tomorrow morning" it was as if I just knew that this will turn out alright in the end."

The following morning I was out of bed by five and feeding the cattle in the dark because I knew we had a long day ahead of us.

"At Seven Oaks Hospital we got her registered and then they sent us to the lab for the tests. There was some trouble at first when they told me that she was only registered for one test but I stuck to my guns until they got things straightened out. I'm sure they are used to dealing with confused old people and they seemed a little annoyed with me at first. This time the old man was right. Thank God that I came along on this trip or she would have been given one scan and sent home.

It didn't take long to take an x-ray of her bad ankle and not much longer to do the bone scan. We were instructed to come back at one-thirty for the CT scan of her blood. It was late afternoon before we left the city to face ninety miles of slippery roads in the midst of a howling Manitoba blizzard. At Lundar the storm was raging full force. It was extremely hard to see the road as we headed west and it was so good to reach home safely. I had to rush to finish chores before dark.

Our experience that afternoon was a good lesson for us that we are going to have to keep our eyes open and our senses about us to avoid getting caught in the shuffle of paperwork."

Now it was a case of "hurry up and wait" until Margo's next appointment. It was terribly disturbing to realize that each day that passed was giving the cancer that much more time to grow and spread.

"January 24, 2003

I phoned the doctor's office today to see if he could move things along regarding her having treatments if it was recommended. The receptionist said the test results are still at the Cancer Clinic but she did move our February 7 appointment ahead by one day. We have to remain patient and positive even if we don't always feel that way."

On January 28, the Cancer Clinic called to say they wanted her there on the fifth of February to arrange treatment. That was music to my ears. Try as I might to remain positive, I'd been secretly afraid that treatments would be ruled out. Now we'd been given some hope to cling to while we waited.

That week dragged by slowly and I spent a lot of time pouring out my feelings in diary letters. On February 6 I found myself typing another letter to Sheila and I remarked that I had finished six ten-page letters in just over a month. It felt so important to me to keep a written record of everything that happened from day to day because all the information we were trying to absorb was overwhelming. It helped ease our fears when we received letters of support coming back to us almost every day.

Here are a few paragraphs regarding the visit with the radiologist on February 5.

"Shortly after noon we arrived at Ernie and Audrey Blue's house where we had lunch. Ernie offered to drop us off at the clinic and pick us up. That saved us the hassle of waiting in line to get into a parking lot as well as a cold walk from car to clinic.

Our radiologist, a lady named Dr. Vijay, had a positive manner about her which we found very reassuring. She seemed shocked to learn Margo had gone through the CT scan on January 17 and we still didn't know the results. The doctor examined Margo's breast and then made a phone call to Seven Oaks Hospital to ask them to fax the CT scan results. We were then left to sit in that stuffy little examining room for half an hour. I had the feeling of a prisoner waiting for the jury to come back in with the verdict.

When the door opened and Dr. Vijay walked in, her bright smile told us we were about to get good news. They don't think the cancer has spread and Margo should be starting her treatments soon. Dr. Vijay will consult with an oncologist to decide which is to come first, chemo or radiation. The radiation treatments are five days a week for five weeks. At least now we have a fighting chance.

The doctor said there is a mark on the lower spine but that could be arthritis. She ordered a couple of x-rays and they were done before we left the building. We were both a lot happier on the way out than when we came in.

We phoned Ernie and he picked us up. We stopped for a few minutes to talk to Audrey before we came home. Those good people have offered to let Margo live with them during the weeks that she is in the city having treatments. That would be a lot better than living alone in the lodge near the hospital."

This was another example of the kindness shown to us when, time after time, people stepped up to make themselves available with no thought of remuneration for themselves.

We got home to find that Ruth had been very busy calving cows all day but with the help of Steven and our neighbour, Garry Seidler, everything was under control.

That night was especially hectic with cows calving one right after another. I was up all night bringing new babies into hot rooms in the barn to get them dried off, so only managed one hour of sleep.

Things began moving quickly for Margo within a few days.

"February 9, 2003

Margo will be going to the city tomorrow to stay with her sister, Kay, for two nights. On Tuesday she'll be having a scan to see if her heart is strong enough to stand chemotherapy. On Wednesday she'll meet with her oncologist to find out what the plans are for treatment. She'll also be getting more blood work done, so things are starting to move at last. Margo has been feeling quite a lot of pain on the left side of her body from her ankle to her shoulder blade, including her groin area. The radiologist said the pain is coming from the cancer and following her nerve system through her body."

That afternoon we left Ruth to watch the cows as we sneaked off to a dance in town. That dance turned out to be great medicine for both of us because of the outpouring of love and support from our friends. It was almost a case of "take a number" if you wanted to talk to Margo and it did her a world of good to know how much people cared for her.

Halfway through the dance Ruth called to tell me that a cow was having a very difficult birth. Einar and Elsie promised to deliver Margo right to our door after the dance was over. I hated to bother them but Margo was happier than I'd seen her in several weeks and I knew she really wanted to stay until the dance ended.

I had a pot of stew waiting on the stove when the dancers arrived and we had a little visit before they headed for home. Kind-hearted Elsie burst into tears when she hugged us goodbye. We told her to quit feeling sad and to keep thinking positive thoughts.

Here is a paragraph written to Donna on Feb 12.

"Mum is in the city this week meeting with her oncologist to find out what course of treatments she will be having. She phoned this afternoon to tell me she has to stay for one extra day of tests. The wonderful news is that the oncologist is very hopeful about her condition. He says she might not have to take chemotherapy, just cancer pills of some kind. I'm not too sure whether I'm keen on that idea but will question it when the time is right. I don't want to take any shortcuts and have her suffer the consequences later."

And the next day.

"Mum left a message saying that she had to stay in the city for another night in hopes of getting in to have an MRI on her liver. The CT scan showed a couple of spots on the liver so they want to check it out to see if it is cancer cells or damage from the cholesterol pills she's been taking for the past five years.

I hate this dirty, rotten, devil cancer that sneaks around undetected until it takes over your body as its own. I hope that all of you gals will be very diligent in checking your own breasts or at least have regular check-ups by a doctor."

Ruth and I were being swamped with a flood of new calves, all arriving during a very cold spell of nasty weather. We were grateful to have Ruth's son, Bob Baxter, visiting for a couple of weeks. Bob helped out at the barn and he also moved the washer and dryer from the basement up to the main floor. Margo was so pleased with the new arrangement that we wondered why we hadn't thought of doing this sooner. This would save her walking up and down stairs with her sore ankle bothering her.

Another journal entry told of our next trip to the city and the horrible news we received.

"We were placed in a small examining room with the door left open so we could see everyone who passed by in the hall. About ten minutes later, a middle-aged couple came walking by heading out of the building. The woman was crying with her face red and streaked with tears. Her husband was stumbling along white-faced and in shock. It reminded me of what we must've looked like on January 2 when we got the bad news from Dr. Haiart. What a cruel world we live in!

A few minutes later the oncologist told us they believe the cancer has spread to Mum's liver already. They are as sure of that fact as they can be without actually doing a biopsy on the two spots there. She also has some fluid around her left lung. We were in total shock and it seemed like the news kept getting worse.

Dr. Brandes told us not to lose hope because he wanted Mum to start right in taking some pills that are used for fighting cancer. I asked him why he wouldn't put her on a course of chemotherapy right away. He gave me a long, level look and said, "Mr. Watson. Why would you want your wife to go on chemo when we know she will get so sick and lose her hair, plus there will be all sorts of other bad side effects? We can give her these wonderful pills and they will do just as good a job?"

I asked him about getting a second opinion and he said that if he had ten other oncologists in the room with us, nine out of ten would be in favour of the plan that he had set up. When he explained it that way, it made a lot of sense. After all, he was the one who had studied for years to gain his knowledge and who was I to question his judgement?

She is to stay on the pills for six weeks and will be re-assessed at the end of March. If there is an improvement, she will continue on those pills and if she is worse, she will likely go on chemotherapy. He said he has had some patients with exactly the same condition that Mum has and some of them are walking around cancer free now.

So there is always that thin thread of hope, that carrot dangling on the end of the stick that keeps us going from day to day. We must trust the ones in charge of her care, hope for the best and find the mental strength to carry on."

As the weeks went by, as always happened during calving season, I started getting very worn out. The nights passed in a blur of cat naps with never more than a two hour sleep between pen checks. It always felt so good to sink into bed and drift away until the alarm rang again. We managed to struggle along and the work that didn't get done one day was always waiting for us the next.

"March 6, 2003

Margo has been on her pills for fifteen days and the side effects all seem to be kicking in at once. She's been having hot flashes and aching joints, as well as headaches and nausea, so her misery continues. She is thankful she isn't taking chemotherapy if these pills are making her feel this bad. I encourage her to come outside and join us but she is feeling much too rough to do that.

Jackie Isfeld phoned Margo last night. She is home after having a complete mastectomy and reconstructive surgery. Too bad they didn't do that last fall but at least now the doctors are quite sure they've gotten all the cancer and she won't even have to undergo any treatments."

Success stories and the constant support of good friends, many of whom are cancer survivors, helped so much to bolster our emotions when we started feeling discouraged. Those special calls always seemed to come just when we needed them most.

We weren't the only ones riding an emotional roller coaster. Ted and Linda had been going through the same thing because their son was battling for his life.

"Darryl had a special test a few days ago and they found no sign of cancer. That is terrific news and it was easy to tell that Ted is very relieved. Darryl has been through a very tough chemotherapy session all winter. If he wasn't young and strong to begin with, the cure would have killed him by now. He told his head nurse he is ashamed of himself because other people in the hospital on chemo seem to be taking it all in stride while most of the time he feels like crap. She told him not to feel bad about that because his course of treatment is the most intensive she has ever seen. Maybe it will all work out in the end for Darryl. He sure deserves some good news after all the bad times he's been through.

Margo is feeling pretty rough most of the time but at least we know she isn't taking sugar-coated candy pills. Last night Ruth took over in the kitchen. Margo has no appetite so hasn't been enjoying cooking like she usually does but she did seem to enjoy eating someone else's cooking."

March 15 brought more good news from Jackie Isfeld.

"Jackie phoned last night with the wonderful news that the biopsies they did on her lymph nodes came back clear of cancer. She has quite a sense of humour and she told me about the reconstructive surgery they performed on her after they removed her breast. They insert an implant and sew the skin up over it. After she is done healing, they will gradually inflate the implant until they have it at the proper size.

Jackie says the darn thing is riding so high on her chest; she told the doctor that she wasn't expecting to get shoulder pads. Then she giggled as only she can. It was wonderful to hear her sounding so upbeat and it did Margo so much good to talk to her.

Margo has been staying in contact with Elsie Kelner who was diagnosed with breast cancer about the same time as Margo. She is in the second week of chemotherapy and is losing her hair. It is wonderful to hear how she seems to depend on Margo for support, calling her, "My rock." Her hubby, Big Al, is getting his hair shaved off in support of Elsie which is something I promised Margo I would do for her if we should reach that stage of this horrible game."

The last few days of March were happy ones because Bertha and Dave had come to visit from the west coast. It was great to see how happy those sisters were to see each other. Dave helped me with chores and we took on the women on a board game called Sequence. Albert and Hulda had built the board for us and had recently taught us to play.

On March 30, we took in a Sunday afternoon dance where we celebrated Bertha's birthday. It was a good way to end the month as Margo was feeling so much better. Things were definitely looking up.

The last day of the month found us in Winnipeg for another CT scan. That day was a real eye opener for us because we were starting to realize that no matter how bad things are, there are always others who are worse off.

"While waiting for the CT scan, we got a good look at a lot of pitiful people struggling to hang on to that last precious thread of life. A steady stream of patients was being wheeled in on gurneys by a couple of overworked orderlies. One fellow was completely yellow so you knew that his liver was shot. Another poor, old guy was lying there with the nurse asking him questions he could hardly understand. He told her he has no family and is all alone. How very sad it is to be alone, especially at a time like that. At least Margo and I still have each other."

The following day we were back in the city where Margo had a blood test before seeing Dr. Brandes. While waiting for the blood test we met up with a young lady from Lundar who was undergoing treatments for cancer.

"We got a surprise when Wendy Sigfusson came running up to talk to Margo. She was in for her first radiation treatment. Last winter she had her breast and several lymph nodes removed, followed by several months of chemotherapy. She's been dealing with the treatments well and looks good in her new wig. She's been a real inspiration to Margo so it was nice to see her there.

Dr. Brandes had a big smile on his face when he entered the examining room, not like last time when he looked deadly serious. He said the scan showed the spots on the liver are stable and there is no sign of new ones. He talked with Margo about the side effects she is having and suggested she take her pill with the evening meal. She lost two and one half pounds in the last six weeks because of loss of appetite. Of course she isn't looking much like Twiggy yet.

We told him we thought the tumour on her breast is getting smaller or softer. He examined her and seemed quite amazed; telling us this was the result he'd been hoping for. I asked him if there is any chance they might operate to remove the spots on her liver. He said that is definitely not an option because the cancer had already traveled from her breast and who knows where else it might be lurking. He said we'll continue to let the pills work their magic."

Needless to say we were ecstatic and almost walking on air on our way back to the car. Margo's face was shining with joy. For the first time we really felt that things were going to work out and that Margo was going to make a full recovery.

With most of the calving done and Margo's health improving we celebrated by taking Dave and Bertha dancing in Eriksdale on the fifth of April. At that dance we heard some good news and some bad—just par for the course lately.

"One of our dancing friends, Ken Thorvardson, told me his wife, Isabel, just found out she has breast cancer. They are like we were at that stage, still kind of in shock but that didn't stop them from dancing up a storm. We all want to make the most of each precious day.

Another friend, Elmer Lutz, was telling me he was operated on for cancer of the pancreas back in 1996. He has recovered and been cancer free ever since. Success stories give Margo and I real hope for a long, happy future together."

Mid-April brought more bad news for Darryl Watson and his family. His cancer was back and plans that had been made for a stem cell transplant had gone out the window. Darryl told me they were down to one last resort—a bone marrow transplant from his sister, Patty. He looked so tired and weak and it broke my heart to think about what may lie in store for him.

On May 12, Margo drove herself to the city for a blood test. Hulda met her in Lundar and went along with her. I stayed home to get the chores done before heading up to Ashern where the weekly cattle sale was taking place. Our herd had grown until we were calving over 180 cows and the work load was becoming unbearable. We had decided to cull our herd and the day before we had loaded thirty cow-calf pairs to take to the sale.

Margo was back from the city in time to go with me to the sale. Together we had the pleasure of seeing our older cows sell for prices that were almost unheard of. The cattle business was booming. The banks had no end of money to lend to any farmer willing to expand his herd. We joked with each other about taking another load to the sale at Inwood the following week.

The next day, I went with Margo to see the oncologist.

"Dr. Brandes talked to us about the blood test results and it was kind of a mixed bag. There are two cancer indicators they watch for and one of them was down from 148 to 100 which is very good news. He was puzzled because the other indicator showed that it was up by quite a bit, which isn't good. He wants to keep Margo on these pills for another six weeks and then do more blood tests and another CT scan. If the results then aren't good, she will likely be switched to chemotherapy or radiation.

The doctor did palpate her liver and it seems to be perfectly normal so that is also good news. The tumour on her breast is getting smaller and we are happy about that."

On May 20 we heard on the radio that a case of BSE or Mad Cow Disease had been found in a cow in Alberta. We'd known for some time this was bound to happen, so we weren't really surprised. That one case caused an upset in our cattle industry that would take years to put right. As soon as word was out, the United States and many other countries closed their borders to all beef imports from Canada. We suddenly found ourselves with a glut of cattle that we couldn't get rid of. Small auction marts closed their doors for the summer until things settled down. Cows that had been worth $1,500 the previous week were now worth less than $100. We were thankful we had down-sized our herd and most of our older cows were gone.

That same evening we received word that another dancing friend, Therese Beliveau, had died from bone marrow cancer. A year earlier she and her mate, Paul, had been at our anniversary party. They had brought us a beautiful, live rose bush which we planted in front of our house. Now, just like that, she was gone. It felt as if all our friends were dying around us. With Margo fighting for her life, it was becoming very important to me that we have something to look forward to after our life here on earth is over.

"Tiny phoned to say that Therese's family had all been called into her hospital room and she wasn't expected to last the night. An hour later we heard she had passed away. I hope she is dancing in her little, high-heeled shoes in heaven now. I wonder if she has met up with her husband, Alphonse, who died from cancer many years ago."

Every time we lost another friend, I did some very deep thinking about what would happen to Margo's spirit if she wasn't able to beat this devil cancer. I wanted so much for her to believe in God in Heaven but how could I expect this from her when I wasn't even sure what I believed in. Ever since Margo had become ill, I'd been praying for her recovery but feeling like a hypocrite all the while I was praying. Several times I tried to talk to Margo about such things but she wanted no part of it. It seemed to her as if I'd lost faith in her recovery if I mentioned Heaven so most of the time I kept quiet and wondered.

The rest of the month passed and Margo was getting out of the house more, planting flowers, mowing grass and helping me with chores by taking twine off the bales when I fed hay. We continued dancing on weekends and it seemed like our life was right back to normal.

June 4 was a happy occasion when we held a cousin party on our deck. Maisie and her husband, Bert, had retired in North Bay, Ontario, and were moving out to Campbell River, BC. Maisie's brother, Jim, had phoned me a month earlier to tell me he wanted to bring them over to meet us. Margo and I cooked up a plan to gather as many cousins as possible to meet this lady who we hadn't seen for many years.

The weather was beautiful and we had about thirty cousins waiting on the deck to surprise Maisie when they arrived. I hadn't seen her since the day she locked me and Ruth out of our own house. It was a wonderful pleasure to meet the gal who had made such a lasting impression on me.

Three days later we welcomed a new grandson into our family when nine pound, five ounce Dylan Richard arrived, the first child of Tracy and Rick. We were ecstatic. Margo was set to take off at once to see him but I convinced her to hold off for a few days in order to give Tracy a chance to recover. This new baby was just what Margo needed to cheer her up. Her health was deteriorating again and even our evening card games couldn't always keep her mind off her problems.

"June 7, 2003

We spent a quiet evening at the Sequence board and were all tied up after eight games. Margo won the last match so that put her in a good mood. She hasn't been feeling very well with pains across her lower back and in her stomach. That has her worried that the cancer is getting worse. I tell her the pain seems to be worse after she eats meat. She doesn't have a gall bladder any more to help her digest food. It's very hard to keep a positive attitude at times like this but we try."

In the last week of June we drove to the city for blood tests at the Cancer Clinic. That evening Margo was given another CT scan to check her liver. Then we spent a very pleasant night with Tracy and her family at their house in Transcona. It was amazing how much the baby had grown since we'd seen him last. We were back at the Cancer Clinic the next afternoon, hoping against hope that we would get good news. I don't think either one of us was expecting the news to be as bad as it was.

"The doctor gently broke the news to us that Margo's cancer is getting worse. The largest spot on her liver has grown a lot and now there are two more spots that weren't there before. I felt like asking him why she had gone twelve weeks without a CT scan but why pick fights with people who are trying their best to help us? He said he hopes to get her on a chemotherapy course as soon as possible along with the usual drug that goes along with it. We can also consider putting her on a new experimental drug which has been shown to prolong life by as much as fifty per cent. The other option is to do nothing and let nature take its course.

The new drug will make her much sicker but greatly increases her chances of surviving for any length of time. She'd have to stay in the hospital overnight after each treatment if they give her that drug so they can help her with the nausea that comes with it.

He gave us some information to read and we are to phone in after we make up our minds. We were in too much shock to think of any questions. I would've liked to ask if there is a good chance of having the cancer go into remission or will it just prolong her suffering? If there is a chance of remission we will choose the more aggressive treatment even if it makes her sicker for awhile.

Neither of us cried on the way home. My tears had been replaced by a red-hot rage at this horrible disease that makes no distinction between good people or bad, rich people or poor. I am mad at everyone and everything—the nurse who checked Margo's breasts last November 6 and found nothing, the mammogram which came back negative, ourselves for going out to BC when we had found the lump the night before. Then there is the whole overloaded health care system which made us wait so long to see the various specialists and also her current doctor who put her on those stupid pills for so many weeks while the cancer progressed.

I'm so ashamed to be feeling this way when I should be down on my knees thanking God for the wonderful life we've shared so far, but I can't help it. Five months ago the whole world was our oyster and we were happily making plans to retire in a few years. Now Margo is sick and our cattle are worth almost nothing. How quickly our lives can change when we least expect it!

At home we had supper before I headed out to get the summer fallow worked. There was a big rain forecast and clouds were coming in from the southwest. Margo got out the lawn mower and by the time I got home from the field, she had cut all the grass. She figured she would feel better doing that rather than sitting in the house feeling sorry for herself. This brave, little lady never ceases to amaze me with her courage and fighting spirit. Somehow, deep inside, I have the feeling that her will to live is going to carry her through this crisis.

We phoned Steven and Tracy and a few close friends but we weren't too much in the mood for talking. We know our friends are anxious to hear the test results but we want to get a grip on our emotions first. I phoned Donna and Kevin this morning.

This evening we got a call back from Kevin. He talked to his mum for awhile and then Nat came on and cried as she talked to Margo. She told Margo that Kevin had been crying all day as they fished. When Margo got off the phone she broke down and really cried for the first time since this whole mess started. As I've said before, she is the strong one in this family and that's good because she's going to need all her strength and more in the next few months."

Now things were happening quickly.

"June 27

I hung around the house yesterday morning until the nurse from the Cancer Clinic phoned. I had left a message there the day before asking if the chemotherapy treatments had a chance of putting her cancer into remission or if it would just prolong her suffering. The nurse said the object of the treatment is to shrink the tumour and give her a better quality of life. Some people have their tumours shrink until they can't even be detected and some of them can live for years. That is good enough for us and I told her we were ready to start the treatments right away. She said she would make all the arrangements and get back to us.

Five minutes later another nurse phoned back and told Margo she is to be in the clinic Monday morning for blood tests. The first treatment will take place that afternoon. We opted for the new drug and she will be kept in the hospital overnight to see how she reacts to it."

The following evening found us at the dance hall in Oak Point taking in our last dance for a long time. We knew that once the chemotherapy started we would be avoiding crowds where Margo might pick up something as simple as a common cold which could be fatal to her. We stayed until almost closing time and Margo was surrounded by people who loved her. Maybe, just maybe, with all those prayers we would get the miracle we needed to save her.

The next day wasn't as optimistic when I wrote about others who were battling cancer.

"This has been a bad week for our friends with cancer. Darryl Watson has been brought out to Eriksdale Hospital where he will spend the last days of his young life. Elsie Kelner is in the hospital with her blood count very low. Margo visited her yesterday and she was in fairly good spirits. Wendy King gave up her fight with brain tumours and passed away on Friday. We seem to be living in a nightmare with no end."

June 30 was to be the big day with the first treatment scheduled. On arriving at the Cancer Clinic, Margo had a blood test and a chest x-ray before we saw the doctor. The first thing he told us was that a urine test done the previous week had shown that Margo had a bladder infection. He wanted to get that cleaned up before she started the treatments.

There was nothing for it but to head back home, stewing about the fact that she'd been allowed to suffer for a week with this infection while her cancer had been given another week to spread, plus it meant another 200 mile round trip. If someone had taken the time to read the results of the tests the week before, she'd have been on antibiotics all week and the chemo could've gone on as scheduled. We were rapidly losing faith in our health system.

A few days later I came down with severe stomach flu. I was just beginning the annual haying operation and for a few days I was much too weak to work. Margo got sick too and she was still sick when we had to head back to the Cancer Clinic the following Monday. This was July 7 which happened to be our forty-first anniversary. We were quite sure the chemo would be cancelled again but to our surprise it was all systems go.

I wrote to Donna on the day following the first chemotherapy treatment.

"July 8, 2:45 PM

Things went as well as expected and the first scary treatment is out of the way. Your brave mum responded like a pro and handled it well. I'd been thinking seriously about quitting my journal for awhile. Now I've come to the conclusion that my words on paper might be the one thing needed to keep all you gals on your toes when it comes to looking after your bodies, especially your breasts. If we'd been more vigilant, things may never have reached this stage and your mum might've been well on the road to recovery right now. The best we can hope for is to shrink the tumours and get her into remission that could last for years.

On the third floor of the Cancer Clinic we spotted some friends, Ron and Shirley King, from Warren. I went over to speak to them and found out that Ron was in for more chemo treatments after his cancer returned. He seemed so relaxed and confident; it was hard to believe he was going through a terrible fight for his very life.

A wonderful nurse named Dawn escorted us through a huge room where people were sitting hooked up to IV lines while the chemo dripped into their veins. Because Mum was having the stronger drug she was taken into a tiny private room where she was able to lie down on a cot. It didn't take long to get the needle into her wrist and the fluid started dripping.

Along with the chemicals she was being given a light sedative to help her relax. The nurse told me to come back about three o'clock when Mum would be taken to the third floor of the Health Science Center.

When I returned, the nurse told me Mum had been awake but was very sick. I could go in but had to be quiet. Mum was kind of out of her head from all the drugs and she got very excited to hear me there. She was asking me all sorts of questions. This new drug gives you nausea which is like motion sickness so it wasn't long before she was up-chucking again. I nursed her through it and then got her to lay back and keep her eyes closed which really helped to ease the nausea. Within minutes she was asleep again. I sat quietly for close to an hour, hoping she would stay sleeping as long as possible.

Dawn came back and started unhooking the IV lines and Mum got sick again. Dawn asked to have the anti-nausea drug, which was supposed to be administered later in the hospital, sent to be put into her by machine. That all took time and it was awful to see this wonderful lady who I have loved for most of my life laying there so helpless, puking her guts out, with her legs and arms jerking and muscles twitching with involuntary spasms. Really lends new meaning to the words, "For better or for worse, in sickness and in health." We know things are going to get much worse before they get better.

The anti-nausea medication settled her down and I was able to help her get dressed and into a wheelchair. A young nurse's aid led us through some dark hallways and through two different elevators.

Once we got Mum in her room, she got sick again before she relaxed enough to go back to sleep. This morning I was back at the hospital before eight and found Mum awake and looking much better. She'd had a wash and some breakfast and was looking forward to getting out into the fresh air. We were home by ten-thirty and our house had never looked so good."

That afternoon Margo did two loads of laundry before collapsing on the couch complaining of weakness. I reminded her that she'd been facing a triple-whammy with the bladder infection, the flu, and then the chemo. The following week was bad with Margo feeling so sick and weak, she was getting discouraged. I was trying my best to get her to eat by cooking special food but it wasn't working. She was still taking antibiotics for the bladder infection and those pills were causing problems.

"July 15, 2003

Margo has been having another really bad day with heartburn flooding her chest. We think the antibiotics are causing stomach upset and she adds to her troubles by spending most of her time lying on the couch. The stomach can't work right in that position but she is too weak and sick to sit up straight. The instructions for the antibiotic say that if it causes stomach upset she can take the pill with a light snack as long as it's not a product like milk.

I hope we get this straightened out. The poor gal has enough misery with chemo treatments without having to battle this urinary tract infection as well. If her blood count drops drastically the infection will hang around for a long time.

We went to Eriksdale Hospital today for Margo's weekly blood test. We did some shopping and then came straight home. I made some nice beef sandwiches for lunch but Margo was too sick to eat. My bullying and coaxing only goes so far and sometimes it's best to leave her alone until she feels better. I spent the next hour trying to catch a nap while screening calls for Margo. She was feeling far too sick to do any gabbing."

That was definitely a low point for both of us but within a few days Margo was feeling better and had bounced back to the extent that she was cooking meals for Ruth and me. It was wonderful to see my sweetheart back on her feet and smiling again. The down side of it was knowing that by the end of the month we'd be back in the Cancer Clinic to repeat the whole nightmare. A routine was being established that would continue for the next ten months. On July 22, I made an entry remarking on a side effect we'd been dreading.

"Margo told me her hair is starting to fall out. I hugged her and told her that I'll love her just as much when she is totally bald. I told her to imagine those chemicals in her blood stream just munching on every cancer cell they can get their teeth into. It's nice to have support when you are sick and Margo knows I am her number one fan. It's good to remain positive and that is easier for her to do now that she is over that first terrible week after her treatment. Now we know what we are facing, it will be easier to deal with next time.

July 26, 6:05 AM

Margo's hair has been coming out in clumps, so she asked me to shave her head. I got out the electric clippers and clipped her hair as close as possible. She looks kind of cute. With her big ears and dark stubble she looks not unlike one of those Tibetan monks that you see in National Geographic."

On August 1, Darryl Watson gave up the fight and died. He'd been through so much in the past year, willing to try everything in search of a cure. He was laid to rest in our cemetery on August 6. Tracy attended the funeral with me while Margo looked after Dylan. I was having serious questions in my mind about why bad things kept happening to good people. I wasn't the only one who might've shared those feelings.

"My cousin, Pastor Bill Watson did a good job as usual but I sensed his faith might've been faltering a little as he tried to make sense of why things like this seem to happen for no good reason. He told us we all have to have faith or what was the point of us all being there and holding a service. I guess even preachers have their moments of doubt."

On August 7, we were back in the city for the second chemo treatment. Margo asked to be put right to sleep while the intravenous was running so she wouldn't have to endure the weird side effects she had experienced the first time. We'd been told to expect such things as hallucinations. During the first treatment she'd seen white doves fluttering around the room and could even hear the noise from their wings.

That evening she was taken up to the same room in the hospital and the nausea was just as bad as the first time. I sat with her for several hours until she finally settled down and then I spent the night with Kay and Alf.

Early the next morning Margo was feeling weak and shaky but otherwise okay. On the way out of the hospital we met up with Al and Elsie Kelner who were in the city for Elsie's pre-op appointment. Elsie was scheduled for her mastectomy on the following Monday and she looked very pale and worried.

This time Margo did so much better. That evening she was up and about making supper like nothing had happened. It was wonderful to hear her speaking in a strong voice. A couple of anti-nausea pills had gotten her through the day and even allowed her to eat a bit of supper. It looked like she might breeze through these treatments and come out smiling in the end.

On August 22, I got a call from John Webb, a CBC television reporter who wanted to talk to us about the drought, the grasshoppers, the Mad Cow crisis and how it was affecting farmers in this area. Buddy Bergner, our auctioneer from Ashern, had given him our names as producers who might be selling our calves at the first sale that Ashern Auction Mart would be holding on September 3. John wanted to make a documentary that would be shown later that fall. He would come up the following week to film at our farm and then meet me on sale day at Ashern. He agreed to give us a day to think it over.

Margo and I talked it over and decided it would give me a chance to say a few things that needed saying. It would also provide the general public a real sense of the disaster the farming community was facing. Margo was willing to go along with it as long as she didn't have to actually take part.

The last week of August found us in the city for a marathon that lasted the better part of four days. Margo was scheduled for a CT scan on Sunday, blood tests on Monday and chemotherapy on Tuesday, providing she was well enough to stand the treatments. We decided to stay overnight with Tracy and her family on Sunday to avoid that long trip back home. This was a bit like mixing business with pleasure. That four day period turned out to be an exercise in pain and misery for Margo and frustration for all of us who were involved.

"For the past month Margo has had lots of pain in her spine and lower stomach. Lately it has been so bad she can hardly walk. Dr. Steyn told Margo she would try to get a spinal CT scan done while we were getting the liver scan done but the technician told us there wasn't any requisition for that one. We got the scan done and then headed out to Transcona where we found Tracy and Rick home. We were both wiped out so we went for a wonderful nap. Rick made a fantastic supper with fresh produce from our farm. We spent the evening spoiling Dylan, who loved all the attention.

Margo was in terrible pain and it looked like she wasn't even going to be able to swing her legs into bed. She finally made it with my help and we fell asleep. The bed was only twin size and it wasn't long before I rolled over and knocked my big, bony knee right on top of Margo's sore hip. She woke up with a scream but neither of us said a word to each other. It was a restless night for both of us.

The next day Margo could hardly walk into the clinic but we got the blood test done and then talked to Dr. Brandes' nurse about the pain Margo is going through. She was very concerned and took a bunch of notes. She was sure the doctor would want to talk to us about it and possibly order an x-ray on Margo's hip. I gave her our cell phone number and she soon phoned to tell us to be back at 1 PM to see the doctor.

We were there in plenty of time but it was close to two hours before the doctor could see us. He is concerned that the medicine that goes along with the chemo is aggravating Margo's arthritis. He may have to take her off that good but powerful drug and we don't want to see that happen. We were both feeling that we should've kept quiet about it but there is no way she can stand the pain on only Tylenol 2. He was also quite upset because all Margo's previous CT scans had been misplaced and there was no way to see what's going on with her liver compared to last month. He sent us to the Health Science building for an x-ray. Margo was in too much pain to walk so we got a wheelchair to make the long trip through the buildings which are starting to feel familiar to us.

The lab was crowded but it was surprising how quickly they got her done. We were given the prints to take back to the doctor and then had another long wait. The other CT scans had been found and the results weren't too bad. The biggest spot on her liver which used to measure about

4.5 centimetres across has now shrunk down to 3.6. The other spots haven't changed and there are no new ones showing. Some of her blood work is showing improvement and some is showing a decline. Her white cell count isn't quite up to snuff for the next chemo treatment. We'll have to make that long drive back to the city tomorrow for another blood test even though there is a good chance we'll be sent home again. We both hate this awful business but there is no choice but to go along with the doctor and hope for the best. We are encouraged with the talk of the tumours shrinking and the one on her chest wall seems to have completely disappeared. Maybe we'll see light at the end of this long, dark tunnel yet. She gained two pounds between these last treatments and that is a good sign.

I picked up a flat of peaches on the way home and Margo made five peach pies.

Aug 27, 8.24 PM

We were on our way to the city by seven o'clock yesterday morning with not enough time to stop for breakfast on the way. Margo got her blood test and the white cell count had come up nicely. Once it starts to come back in the right direction, it really comes fast. By eleven o'clock Margo was being hooked up to the IV machines and I left her there.

Margo was sleeping peacefully when I got back but she woke up while we were moving her upstairs. The motion of the gurney on the rough floors made her sick and she spent the first half hour in her room throwing up. It was incredibly hard to see her shivering with chills one minute and then burning up with fever the next. All the while that horrible, metallic smell of chemicals was radiating from her every pore. We are starting to find out why we'd been told that sometimes the treatment is worse than the disease."

On August 29, John Webb came out by himself to do the interview. John would be running the camera as well as asking the questions. We spent a long afternoon driving here and there on the farm looking at the effects the drought was having on our land. He shot hours of tape but told me he'd be lucky if we ended up with a twelve minute segment.

We also stopped in Eriksdale and Lundar to interview businessmen to get their opinions on how the BSE crisis had affected their business. Then it was back to our deck where we spent the last hour before dark talking about our plans for the future. I told him that for now the farm was on hold as we dealt with Margo's illness. Her fight for survival was first and foremost in our minds, as it had been for the last ten months.

On September 3 our calves sold at Ashern in the first sale held there since the Mad Cow had been discovered on May 20. John was there to film our calves being sold for much higher prices than we'd been hoping for. After the sale, he followed us home and did another interview on our deck, regarding our feelings about our decision to sell early. He said he'd be back in about a month to wrap things up.

That week we made a decision not to attend the annual corn roast at Les and Margaret Hayward's farm near Lundar. We'd been going there without fail every year since they built an outdoor dance floor about ten years earlier. We always enjoyed the music, dancing, and best of all the huge pot luck supper. This year we didn't dare take a chance of having Margo come into contact with any flu bugs while her immune system was so low.

On September 6, two nice things happened. The United States opened its borders for boneless beef from Canada. Our feeder prices immediately improved which meant that we made the wrong move by selling in the first sale. On the other hand, our cows were doing much better now that they didn't have those big calves dragging them down.

That evening Margo was feeling so much better we had Larsons and Bergs down for supper followed by fourteen games of Sequence. Ray and Georgina hadn't played before but they caught on quickly. We had a wonderful fun-filled evening together. With all the laughter and cheeky talk flowing, it was hard to believe there was anything wrong in our world.

September 16 brought another chemo treatment for Margo and she handled it like a pro even though she was dreading it so much she got sick on the way to the city. I spent the night at Tracy's house while Margo was kept in the hospital. Early the next morning we drove home to find two inches of rain had fallen while we were gone. This was wonderful news as it assured us we would have some fall pasture and moisture in the ground for next year's crop. The rain continued off and on for another week until it reached a point where we were noticing hay bales sitting in water in the bottom of some sloughs.

I'd been praying for help and lately it seemed that God was starting to answer my prayers. Of course He had been answering them all the time but I was too preoccupied to listen and look.

"September 24, 2003

I was tired last night and fell asleep quickly, only to become wide awake at two in the morning. It was so bright in the room it seemed as if it was breaking daylight. I walked out on our deck and was greeted by a dazzling display of northern lights. Do you all remember a few years ago when I wrote about seeing a display of lights one night? That display was like a message to me that there really is a God and Heaven.

Last night I looked up and prayed to God, asking Him to save Margo's life. I refrained from trying to make any bargains with Him or any promises that I might not be able to keep. I felt a little bit better when I came back in.

We certainly are in great need of help here, more than modern science can give us, I'm afraid.

Margo's pain is getting worse every day. It is so hard to see her suffering and to feel so helpless to do anything to comfort her. All I can do is support her and pray for her recovery. Miracles do happen!"

As strange as it may seem, this was also a very happy time for us and the good times outweighed the bad. Dot visited for a couple of weeks, sharing her time between our house and Ruth's place. She was with Margo when I was out hunting geese with Ron Heroux and one evening she treated us to supper in town. Margo had been cooking again whenever she felt up to it. The night out gave her a chance to sit down and be served by someone else.

"For the past three days Margo has been feeling so much better, she is almost like her old self again. The night before last we were playing Sequence with the sisters and I never heard her laugh so much all summer. Her voice is strong again and she is cheeky and full of spunk like she used to be. Her appetite is much improved and she is eating well again. I'm taking this all as a good sign that her cancer is taking a beating from the chemo treatments. We are looking forward to and also dreading the results from the next CT scan in about nine days."

The following week we were told that her cancer was basically staying stable and she was given another chemotherapy treatment which she handled quite well.

We were feeling more at ease at the clinic and I noticed Margo making real efforts to reach out and talk to people who were obviously beginning their own trial by cancer. Margo had always been shy and avoided speaking to strangers so it came as a bit of a surprise to me to see her making the first move. She said it made her feel better when she was able to make someone else feel more at ease.

All summer we'd stayed away from dances for fear of Margo catching a cold. We were missing our friends so much that we finally took a chance and started going out again.

"October 12, 2003

Elsie had phoned Margo wanting to know if we were going to the dance in town. As the evening passed, Margo kept hinting about it until I finally asked her straight out if she wanted to go. She said she would love to so I hurried to get ready, and then helped her get her wig on straight and we made it to the dance about an hour late.

It was so worthwhile when we saw how happy our friends were to see us. There was a line forming at the end of our table with friends wanting to talk to Margo and wish her well. Margo had a few dances and we went home a bit early but very happy.

We are told that Margo's mental outlook is very important in fighting this disease. There isn't much sense in staying at home becoming depressed when we could be getting out once in awhile. From now on we'll try to work out a balancing act. The hardest part is quickly telling people not to hug Margo. They all want to hold her hand which is just as bad for transferring germs. Last week, one of her gal friends gave her a big kiss just before telling us she had a horrible cold. People mean well but sometimes they just don't stop to think."

October 21 brought a letter from Heather Smith Thomas from Salmon, Idaho. In spite of never having met her in person we felt very close to her and looked forward to a day when we would meet her and her husband, Lynn. She'd written a beautiful letter of support, one of many that she would write in the months to come. We knew how busy she was with her writing and ranching so we truly appreciated the time she took to lend support to us.

Late in October we attended a supper dance at the Oak Point hall which was where we had fallen in love with dancing again. Once again, Margo was swamped with people coming to visit and wish her well. She looked pretty cute all dolled up. With her wig and make-up, she appeared quite different than she had before she got sick.

Next day's mail brought another letter and small package from Heather Smith Thomas.

"The package contained a book written by a young lady named Renee Bondi who had broken her neck in two places when she tumbled out of bed one night. She became a paraplegic and her whole life changed. The book is called 'The Last Dance but Not the Last Song.' There was a long letter of encouragement from Heather and she sent the book to help us stay positive.

Heather's faith has grown very strong through all of her own troubles and I envy her that faith. I'm very much like my dad was when he was about the age that I am now. He wanted to believe in God and possibly, in his own way, he did believe. I also want to believe and sometimes reach out in a tentative fashion to people who have strong faith. Then, when they come on too strong in their attempts to make me a believer, I find myself pulling back, not wanting to be preached to or at. I feel it would be easier to accept what we are going through if only my faith was stronger. Maybe Renee's book will help us both. Margo is reading it already."

The last few days of October saw Margo go through another chemotherapy treatment. Our Manitoba climate brought a change in plans.

"October 30

We've been in the city for three days straight. We went in Monday morning planning on having Margo's blood tests done before coming home again. We got caught with the first bad snowstorm of the winter so decided to stay in where we were safe. We didn't want to miss out on Margo's sixth chemo treatment the next day.

The next morning Margo's blood work was repeated and then we saw the doctor. Her cancer is staying stable and hasn't grown or shrunk since the treatments started. He'd like to give her another treatment and then try a different drug with a series of four treatments. She is unable to stay on the first drug for more than six treatments for fear of damaging her heart.

We'd been hoping for better news but it could've been worse. He might've told us there would be no more treatments and that we should go home and get our affairs in order. At least we've had more time together and her cancer isn't any worse than it was when the treatment started. By eleven o'clock she was being hooked up to the IV line.

By two o'clock, Margo had finished her treatment and was sound asleep. The nurse said they would take her to her room in the Health Science Centre about four. I leaned my chair back against the wall and slept for awhile.

This time we put her in a wheelchair to move her. She made the trip without getting sick and we were grateful for that. I helped her into the bathroom for a pee and she was safely tucked into bed by the time the nurse showed up to help. This was the first time she made it through a treatment without getting sick and that was a major relief for both of us. I sat and read while she slept just so she would see me each time she woke up. She tells me it means a lot to her to have me with her."

On November 3, I got a call from the CBC reporter. After a bit of coaxing from me, Margo agreed to take part in the final episode of the CBC documentary.

"John wanted to end the story with us talking about Margo's cancer and her chemo treatments. I told him we would talk it over and let him know. He said he could come out here and finish it off with me cooking a meal for Margo to show how our life has changed this summer. I do feel it would be a great film and I'd like to see it finished."

"Yesterday morning found Margo feeling kind of nervous about the filming. She told me straight out she'd been talking to Steven and he'd told her not to let me push her around. I guess I've been pushing her a bit on this as I've always done. In my own defence I have to say that she too has always had ways to get what she wants when she wants it. She is a little more subtle than I am and prefers to work slowly and use the old "wear me down" approach. I was quick to reassure her she wouldn't have to do anything she wasn't comfortable with.

I had to hurry out and get my cattle fed before John arrived. This time he brought a camera man with him, Jim Neilson, who seems to be a super guy. Jim told us his mother has cancer on her large intestine and has recently finished six months of chemotherapy.

We talked about how we were going to do this last segment. Jim had to fool around experimenting with the lighting before we started filming with me making lunch for Margo. I did the whole meal, peeling spuds and talking a bit. Once I had the vegetables cooking, I called Margo and asked her if she wanted to get her butt kicked in Sequence. We had time for one game while the spuds boiled. Then she made her escape to the living room while I fried the pork chops which fortunately, I had remembered to take out of the freezer to thaw.

Jim filmed us having lunch and once he got what he wanted I set two more plates and they had lunch with us. They were super grateful to be invited to eat.

I got things put away and then Jim shot John interviewing me about our decision to sell our calves when we did and my thoughts on the present state of our cattle industry. He had lots of questions regarding how our plans for retirement have changed with the BSE crisis as well as with Margo's illness. I told him our main concern is having Margo get better because we'll always find ways to cope with the other problems.

The documentary is going to be on Canada Now on CBC evening news, hopefully on November 17. It might be shown on the National if it's good enough. John has a ton of work to do before then, editing the many hours of tape into one short segment.

By the time the guys left, Margo had warmed up to them and she was talking a blue streak as long as the camera wasn't on. She's such a shy lady."

Margo spent her birthday in town getting some work done on the truck. Later she ate lunch with Adeline Backman in Lundar, followed by a doctor's appointment in Eriksdale. That wasn't too bad for a gal in the middle of a long session of chemo treatments. I sat at home taking phone calls for her that came in non-stop. It was exactly one year since the mammogram which had pointed to her problem. It was hard to believe how our life had changed in one short year. We'd gone from planning an enjoyable retirement to trying to get through one treatment and moving onto the next in hope of having her enter remission.

November 17 found us back in the city for blood work and another CT scan. We spent the night with Tracy and watched for our documentary but it never came on. It turned out they were swamped with high priority news stories and our bit was postponed.

We had a great supper and a good sleep. We were back at the clinic the next morning for more blood work. Her blood count was border line for her chemo treatment but it was done anyway to avoid getting behind in the schedule. There was a whole pile of papers for her to sign before starting the new drug. After she was hooked up to the IV lines I headed home. Winter had arrived which meant I had to feed cattle and fill the furnace.

Here's a paragraph dated November 25 regarding Margo's reaction to the new drug.

"Margo has done very well in the past week since receiving the new drug along with the chemo. Yesterday was a bad one and she felt sick enough to throw up. One of the side effects of this new treatment is that it might cause her to lose her finger and toe nails. Her nails are really sore already so we are prepared for that to happen. This is more torture to add to her misery."

On November 25 our documentary was finally shown in the evening news.

"John Webb phoned to say the show would be on at 6:15 PM. We made a few phone calls to friends and then settled down to watch. John had done a great job of editing in the five minutes he had to work with. The whole documentary lasted about ten minutes and we shared the time with the Green family from Fisher Branch.

We were really surprised to see a few seconds of Harold Backman and me playing music in our kitchen. John had told me he was determined to squeeze a bit of our music in. There were also a couple photos of Margo and me dancing at our anniversary party last year."

After the show aired, our phone never quit ringing for days. Our brief moment under the spotlight brought more attention than we'd been looking for. Margo was experiencing serious side effects from this new drug. We were also quite concerned that her cancer was advancing.

"Margo is still having all sorts of miserable side effects from the drugs and her stomach is hurting her constantly. Yesterday she had an appointment with Dr. Steyn in Eriksdale. She came out with a requisition for some supplies to do some home tests. The doctor said the pain isn't coming from her liver but from her stomach. The chemo treatments are making our lives hell. The up side is that we are both sure the tumour on her breast is fading away to nothing. Maybe the same will happen to the ones on her liver.

I told Margo that if she is feeling this bad she should imagine how bad her cancer must be feeling. Most likely it is almost dead now and ready to go into remission. She smiled and told me that I'm her number one supporter."

On the sixth of December we attended the annual supper and dance at the Lundar Legion where Margo and I had been members for many years. The meal was very good and the dance was great though not as crowded as usual. Margo's bad hip didn't let her dance much but she enjoyed the music and visiting with friends.

We had no way of knowing what the next few days would bring.

SEVEN
My Return to God

The morning of December 8, 2003 found us on the highway to the city well before daylight. The moon was full and bright and we could see deer grazing in the fields. Margo had her blood test done at the Cancer Clinic and we did a bit of visiting and shopping before starting the long drive back home.

The next morning we returned to the city where we learned that Margo's blood count was well above acceptable limits. This allowed the chemo treatment to go on as planned. The doctor said the tests showed her tumours had shrunk forty per cent since the last treatment. The new drug was working. This was the first real good news we'd received in over a year and it had come just in time for Christmas. The smile on Margo's face was as bright as if she had just won a lottery and indeed she had—the lottery of life.

I left Margo at the Clinic having her treatment and returned home to tend the cattle, promising to be back early the next morning to pick her up. That evening I was stricken with an ailment that drastically changed our lives and our plans for the future.

"I wasn't feeling very hungry but figured I'd better eat something. Supper was a venison-pork sausage and a dried-up jelly doughnut. An hour later I started getting terrible pains in my stomach which gradually got worse as the evening went on. My stomach was bloated and the pain was almost unbearable. I finally went to bed and slept fairly well but the pain was still there this morning.

Dec 11, 8:45 AM

I was on my way to the city before six o'clock yesterday morning to pick up Margo. It had snowed a bit in the night and the highways were very slippery. I got to the hospital half an hour late and Margo was becoming very worried about me.

Last night Margo spent hours on the phone telling people her good news while I lay around on the couch wondering what in the world was going on with me. The belly ache continued all evening and got worse after I went to bed. By midnight I was wide awake and fairly sure I had something nasty like an ulcer or stomach cancer.

This morning Dr. Watson has diagnosed his own ailment as a gall bladder attack. I don't know why it took me so long to come up with a logical solution. The pain is high up under my ribs and Margo tells me that is where she had pain when she had the same thing a few years ago. The pain is mostly gone right now but my liver area is very tender and sore.

Dec 12, 7:05 AM

I wish I could say I'm feeling better this morning but that isn't the case. Sleep has been hard to come by and I've only been dozing since about two-thirty. The pain is lower in my belly now and even Tylenol doesn't seem to be working. This is really starting to worry me.

Dec 13, 7 AM

Yesterday Margo did some phoning around trying to find a doctor who would check me over. Dr. Burnett in Lundar managed to squeeze me in at about eleven in the morning. I was given blood tests and a urine test. The urine test was okay but the results of the blood tests will take several days to return.

That afternoon I was feeling like I'd been run over by a herd of wild horses. My bones and joints were aching and a strange feverish feeling was running through my body. Whatever had been wrong with me earlier in the week seemed to be turning into a severe case of flu. Unfortunately, I had already swallowed a cup of prune juice and a tablespoon of castor oil.

We'd been planning on going in to the Elks supper in Lundar but I told Margo there was no way I could attend in the shape I was in. She got ready and went by herself, meeting Ruth at the hall. I managed to catch some sleep on the couch between wild dashes to the can. Warning! Don't take prune juice and castor oil when you are coming down with the flu. You might get much more than you bargained for.

Margo got home and went straight to bed. I convinced her to sleep in the other room so she wouldn't be disturbed. This would also keep her away from my germs. I fell asleep on the couch and woke up feeling thoroughly chilled before I scurried to my lonely bed with teeth chattering like castanets."

On the morning of December 16, Margo convinced me to go to the emergency room at the Eriksdale Hospital. She wasn't getting much sleep listening to me rambling around the house all night. The final straw came when I spilled a full jug of milk on the kitchen floor at three in the morning. When she got that look in her eye I knew better than to resist.

Here are some entries from a letter to Donna.

"I've been in Eriksdale hospital for the past three days. The pain is gnawing at my guts and I still have to wait for another hour before receiving my next shot of painkiller. Writing might help take my mind off my pain.

Early Tuesday morning I caught a ride to the hospital with Mum as she went in for her weekly blood tests. The nurses settled me in at nine o'clock on a narrow, hard gurney in an examining room. They phoned Dr. Burnett and after he examined me I was hooked up with an intravenous drip. Demerol chased some of the pain away but Gravol made me really drowsy. I was running a fever and received Tylenol for that.

Yesterday morning I was informed there would be no more food or drink for me. The only time I get a drink is when I take a pill and then it's only a sip. In the afternoon I got the good word that I am going to go by ambulance to Selkirk for tests tomorrow. I was hoping it would be today.

Later, same day, 11:30 PM

It's only been 3.5 hours since my last fix but the stabbing pain in my gut has me awake already. The doctor had left word that I was to get Demerol every six hours but today my day nurse phoned the doctor and he dropped it to every four. They tell me I'm getting a very small dose so I was happy to hear that.

Today was pretty busy so the time passed fairly quickly. Ruth stopped in after her hour at the post office to bring me some mail and this writing paper. Mum came in after lunch for her appointment with Dr. Steyn. You should see her poor feet. Her toes are reddish purple and hurt constantly especially when she's walking around. The good news was that we were expecting Tracy and Dylan out for a visit. I was very grateful for that because I hated to leave Mum home alone.

A few minutes later, Garry and Corrine dropped in for a visit. I could see the fear in his eyes as he looked at me. Ever since they lost their son to cancer a few years ago, Garry hates to hear about anyone being sick.

Norman Kirby came in this evening. He was here visiting his brother Arnold and is pretty worried about him. Arnold has been throwing up almost non-stop for over a year now and the doctors can't seem to find the cause. He's been in here for ten days, hooked up to an IV line.

Finally all was quiet and I slept well until the pains came back about eleven. I'm past the four hour limit now so will call the nurse for that wonderful pain killer. Good night, I love you all.

Dec 20th, 7:20 AM

Yesterday I was taken by ambulance for tests at Selkirk Hospital well fortified with a shot of Demerol to control my pain. By the time we reached the rough stretch of highway between Clarkleigh and St. Laurent I was feeling quite happy. Every time we hit another pot hole, it was just "Bump bang, let her buck, boys—who gives a hoot?"

At Selkirk I was given an ultra sound of my gall bladder and a barium x-ray. The ride home wasn't as pleasant because I was coming off my high and the pain was returning full force.

Sunday Dec. 21, 5:35 PM

The nightmare continues with no sign of improvement. The Demerol relief only seems to last two and a half hours, followed by an hour and a half of sheer agony. I'm no wimp but this is hard stuff to take. Sometimes it makes me feel like bursting into tears or else tearing off my IV lines and running screaming down the hall. Each attack lasts until the next shot of Demerol kicks in. If this is a small sample of what Hell is like, I'd better get busy and mend my fences with my Creator.

Steven popped in this morning on his way to the farm. He's going to take over the feeding to give Garry Seidler a break. Steven offered to quit his job for awhile and look after the cows during calving season for me. What a good guy!

Mum and I and the doctor on call discussed moving me to the city tomorrow morning.

11:30 PM

My Demerol dose has been increased. The doctor came back and said he'd been checking my records. A blood test for the pancreas had come back clear. That was real good news."

I was getting quite worried about the whole situation. The way things had been going it seemed to be a downhill ride. What was really worrying me was what kind of Hell I was going to find at the end of that ride if I didn't come out of this alive.

Here is part of a letter that was started several days later.

"Victoria General Hospital, Dec. 24, 10:45 AM

This isn't my idea of the perfect way to spend Christmas Eve but I'm content to be right where I am. At least they can keep my pain under control as they search for the cause of my problems.

On Monday morning I went to the city by ambulance right after lunch. My nurse gave me a light dose of Demerol to kill the pain on the trip down. That struck me as rather strange because I knew they'd given me a heavy dose for the trip to Selkirk last Friday.

By the time we arrived at the hospital, my pain was coming back. I was wheeled down a hallway filled with sick people on stretchers, straight into treatment room #9. I was feeling a little guilty because I thought I was going to be admitted right away while those other poor souls looked like they had a long evening ahead of them. The whole admitting area was bedlam and I watched through the open door as more stretchers arrived. The pain in my stomach was now full blown and radiating right around my back.

The two big muscles along my spine were as tight as fiddle strings and my fever was spiking again.

At five o'clock a nurse came in and told me she'd phoned Dr. Cowan and he was on his way over. That was the most wonderful news. I was still on a gurney hooked up with my IV but I was used to maneuvering with it by now. I unplugged the machine and dragged the pole with me as I headed over to a washroom. Then it was back to the torture chamber where I tried every kind of position known to man and some that aren't. I would try not to look at my watch until I was sure half an hour had gone by and then would be amazed that it was only three minutes.

Six o'clock came and went and still no doctor. I wasn't having the kind of pain that lays you out, but the kind that just keeps eating away and wearing you down. The only thing keeping me going was the hope that the door would open and the doctor come in.

At seven o'clock a nurse came in with a message from Dr. Cowan saying he wasn't going to arrive until later in the evening. I was stunned and stared at her in amazement. I asked if she could at least give me some Tylenol but she curtly replied that it was against the rules. Now I really entered panic mode and of course my anxiety made me more upset and that only aggravated the pain.

About five to eight I heard the door open and sat up expecting to see the doctor. Instead it was Steven and Linda, who had been in the city shopping. As soon as I saw them I broke down. The floodgates burst and I cried like a baby. I was feverish and sweaty and must've stunk like an old dog. Dear Steven sat beside me and rubbed my sore back and told me I was going to get through this. Linda stood like a stone statue gazing at me with red eyes and tears pouring down her cheeks. Talking to them kept me calm until the doctor finally got here at nine. Dr. Cowan asked Steven to wait until he'd examined me. He was probably hoping to send me home again but it would've taken dynamite to blast me out of here.

This doctor had looked at my records and, like the doctors at Eriksdale, was mystified as to what might be the problem. He said my high fever indicates an infection but all the blood tests are normal. He admitted me with orders for shot of morphine to dull my pain. A nurse ran the life saving fluid in and five minutes later I was asleep."

I was scheduled for a CT scan the next day but it was cancelled because my stomach was still lined with chalk from the barium x-ray. A prescription for a strong laxative was added to my files. I was given the laxative about ten minutes after a morphine dose and the stomach cramps hit before the narcotic kicked in. I was in for a nasty night.

"Dec 25, 5:30 AM

Merry Christmas, I love you all and prayed for all of you last night.

Night before last was pretty wild. I would doze off, only to be awakened by cramps that sent me rushing to the bathroom pushing my IV pole ahead of me. I'd asked my nurse to cut down on my morphine during the night. I didn't want to add more problems to existing ones by becoming addicted to morphine. This nurse, Annie, was doing a double shift, 16 hours. She can be proud of herself. Her daughter is a journalist and her son is a doctor.

Jonathon, my day nurse for the past two days, is a really super young guy. He disconnected my IV and wrapped my arm in plastic to keep the water away from all the tape holding the needle into the back of my hand. I had a long hot shower and could feel the tension leaving the sore muscles in my back.

Yesterday morning they took me down for a stomach x-ray to see if the chalk was gone. It was all clear, which wasn't surprising. There was no mention of a CT scan though.

All morning I kept feeling better and was bright-eyed and bushy-tailed when Tracy and Dylan arrived about noon. It was so good to see them and get to play with my adorable little grandson. He is so cute and cuddly.

About one o'clock, Dr. Cowan arrived. He'd received the wayward Barium x-ray from Selkirk which showed no ulcer. That's why he cancelled the CT scan. He's still completely mystified but talked at length about my condition. He wanted me to try switching from morphine to Tylenol when I get pain. I told him I hadn't had a shot of morphine for seven hours and was feeling fine.

The doctor told me to try eating real food so Jonathon brought in my tray that was still on the cart in the hallway. I ate half a bowl of vegetable soup before my pain returned full force. Within ten minutes I was rolling around in agony and sweating bullets. The nurse brought Tylenol and I sent Tracy and the boy home. I woke up in an hour with the pain gone.

The rest of the evening was quite good and I walked the long hallway from one end of the hospital to the other for more than 35 minutes, pushing my IV pole along with me. Supper didn't bother my stomach.

About eleven PM, I had another attack coming on. I tried to tough it out on Tylenol but gave up at 12:15 AM and got a shot of morphine. Another attack this morning but a very small dose of morphine soon set me right. Love you all, Dad"

Christmas came and went. Dr. Cowan told me I would have to be half-dead before he could get any tests done because most of the technicians were gone for the holidays.

On Christmas Day I received a wonderful letter from Heather Smith Thomas. In spite of all the trials her family had been through, her faith in God never wavered. Before starting another diary letter, I wrote to Heather, bringing her up to date on what had been happening with us. I mentioned I'd been struggling with my faith, or lack of it, since my mum died.

"Although I was raised with plenty of religious training and was baptized and confirmed in the Anglican Church, I've never honestly had complete faith in God. Looking back at all the close calls and near-death experiences in my life, I've come to have a firm belief that I have a guardian angel who works overtime to keep me safe. With all we've been through lately, I've decided it is well past time to call my angel by his proper name, God.

I never tire of seeing new calves born and never fail to marvel when that little creature takes its first gasp of air and opens its eyes for the first time. At times like that I know in my heart this has to be more than a simple accident of nature.

Margo has no faith in God at all, which grieves me deeply. Our parents were such good people; I really believe they are walking in Heaven. I'd like to see them all again some day with my dear Margo walking by my side."

The last few days of the year dragged by and on December 29, I started a new letter.

"Dear Dot,

I'm still in a state of limbo at Victoria General but there is light at the end of the tunnel. My doctor told me yesterday there is a good chance there will be some tests done today. For the past four nights I've been having pain attacks that wake me up after a couple hours. A small shot of morphine sets things straight and I can go back to sleep. During the day, while I'm upright, I'm fit as a fiddle.

In spite of being miles from home, I'm getting lots of visitors. The night before last, Ernie Blue brought in Robert and Jackie Isfeld. They brought a guitar and fiddle and we made music for two hours straight. That did lots to boost my morale but the excitement brought on an attack that night that had me shivering under five blankets. Yesterday was Sunday and despite bad weather, I got quite a few visitors and lots of phone calls. Some of my friends sound hesitant at first but relax when they hear me sounding happy.

My stay here has been a real eye opener for me. Because my doctor thought he'd be operating on me soon, I found myself on the fourth floor which is the post-op ward. Talk about pain and suffering! I speak to any and all who are lonely and scared and in pain. I seem to have formed friendships with the special ones. Most will likely be ships that pass in the night but there's a couple that may prove to be long lasting.

There is Bill Beggs who had prostate surgery three weeks ago followed by gall bladder surgery last week. We met in the hallway one day while I was walking with "Ivy", my tall, skinny girl friend. He was taking his first hesitant steps with his wife, Lorraine holding his arm and pushing his own IV pole. I spoke first when I challenged him to a foot race to which he replied, "Not today, my friend." Over the next few days he grew stronger and we did lots of talking or maybe I was doing all the talking while he listened. One day he looked at me and told me I shouldn't worry so much because it wasn't doing me any good. I asked him how it was possible not to worry about BSE in our cattle, Margo sick and undergoing treatments for cancer, and now my illness.

Bill replied that we are all just a tiny part of the Master's plan. He suggested I relax and put my faith in God and I might be surprised how well everything would work out. Those few words spoken so quietly had an immense effect on me. I was rapidly approaching the point where I realized I wasn't as strong and invincible as I had believed. Bill is being discharged today and they are moving to Kelowna, BC next week.

There is also 95-year-old Leonora who is recovering from hip surgery. She told me the list of ailments she's suffered all her life. As a young woman she contracted tuberculosis and spent five years in the Sanatorium at Ninette, MB. Part of the treatment was collapsing one lung at a time to let it rest for a couple of weeks. Another part was being bundled up to sleep out on the veranda at thirty below F.

I was amazed to learn she's been a doctor all her life. In fact she'd just started her practice when she caught tuberculosis from one of her patients. After she recovered she went right back to her practice and served faithfully for the rest of her working years. In the mid-nineties she was diagnosed with cancer in her breast and it was removed. A few years later, she had the other breast removed. Now she has the broken hip to contend with but she's a brave old lady and a survivor.

Dec. 30 1PM

Yesterday Tracy and Margo visited when Margo had several hours to spare between her blood tests and her CT scan. They brought my guitar and later on, I played a few tunes for my roommate, Ken. Manipulating my wrist caused my IV site to leak fluids. I'd finished my last bag of antibiotics so they were able to take the needle out and leave it out, at least up to this point.

After a pleasant evening, I asked for two Tylenol to ward off the nightly pain attack. It seemed to work and I settled down for the best sleep in three weeks. Sometime in the middle of the night, I awoke out of a sound sleep, suddenly aware of a presence standing near me. In the glow of the streetlights, I could see a beautiful woman gazing at me. I was so startled I gave a violent jerk.

My first thought was that it was a ghost but she was so beautiful and looking at me so gently, I just knew it was my guardian angel. I've always felt that my angel is a man but this was definitely a woman. She looked middle-aged and had long white hair. She was dressed in a beautiful white dress. My next thought was that I had died and she'd come to take me home. Before I could speak, she asked me in a voice so soft it was barely a whisper, "Are you okay?" I replied, "Yes." She turned and walked softly out of the room. I wanted to call her back and ask her the meaning of this but she was gone.

I sat on the edge of the bed and tried to make sense of what I'd just seen. Was it a dream? No, it had been far too real for that. Was it my night nurse checking on me for pain? No, that nurse was young, dark-haired and wearing a dark sweater. The only logical explanation was that my angel had come to check on me and for some reason had decided to reveal herself to me.

By six o'clock, I was wide awake and excited about what had happened. I washed my face, brushed my teeth, had a shave, and took off for my morning walk. Part way down the hall I found my angel sitting at a nurse's station checking records. There was no mistaking her so I approached and asked if she'd been in my room in the night. She said she had been, so I told her at first I'd thought she was a ghost. She said she'd seen me jump and apologized for scaring me. We shared a good laugh over that. She'd been filling in while my night nurse was on a break.

This was a strange event, but kind of fun too. It confirmed my faith that I have absolutely no fear of death. I look on death as another great adventure that will end with me meeting up with friends and loved ones who have gone ahead. The thing I fear most is ending up old and infirm, strapped down in some institution for months or years.

Later I was taken down for a special CT scan which will show whether there is a blockage in my bile ducts. I had to lie on my back for a full hour while the machine took constant scans of my liver. It felt rather strange to see my liver displayed on a screen.

A few minutes later I was given a bottle of pure cream to drink and sent back to my room. There was another scan done an hour later. I'm so happy to be finally getting some real action. Now that I'm feeling better, I'm chomping at the bit to get back home again.

Between scans I phoned Tracy and she gave me wonderful news. Margo was in for a chemotherapy treatment and her doctor said her tumours have shrunk another 50%. To my way of thinking, that means her tumours are only 30% as big as they were six weeks ago. I've been praying for news like this. When I saw the sun rise in the blue skies this morning, I had the most positive feeling for both of us.

Dec 31, 2003 evening

Happy New Years to everybody lucky enough to be out in open air."

The long sad year had come to an end leaving me with a mixture of emotions. The bad news was that I was no closer to finding out the cause of my mysterious ailment than I had been two weeks earlier. The good news was Margo's cancer markers were coming down. It seemed as if the chemo was working and we had real hope for a long happy future together.

My time spent in Victoria General hadn't been wasted. I met many wonderful people, most of them suffering intense pain after their operations. I noticed the people who had a very strong faith in God and eternal life seemed better able to handle their pain. They were calm and serene as they put their faith in Him. I went to bed on New Year's Eve hoping to sleep the night through but it wasn't meant to be. God had big plans for me.

"Jan 1, 2004 12:45 AM

The hands on the clock ushered in the New Year about three quarters of an hour ago. My New Year came in with a fresh blast of pain about ten minutes past the hour. I went out in the hall and found my nurse. She gave me Tylenol before I went back to bed but the sharp waves of pain kept me awake. I'm now sitting in the cool room at the south end of the hall with my blanket draped around me.

The last day of the year was wonderful, starting off with a two hour early morning visit from Steven and Margo. He'd braved the icy highways to pick her up after her chemo treatment. At the Cancer Clinic, a lady had presented her with a small afghan. It is machine sewn but warm and beautiful and she totally loves it.

Margo and Steven left about noon. We were standing near the elevators when the door opened and out walked my cousin, Eunice, who has been so faithful to me. She was so happy to see Margo and they shared a big hug. Eunice and I spent half an hour in the television room. I filled her in on my latest happenings and we discussed the dismal state of our health care system.

Tracy and Dylan arrived after lunch to find me sleeping and we had a great afternoon together. While they were here, a nurse brought a request from Leonora and Margarita in 426. They wanted to hear me play and sing. I did so right away and that brought more requests for me to sing for other people who are really sick and in pain.

The evening brought a great surprise when Phil and Rose Leitold from Woodlands dropped in to visit me. Rose is a cancer survivor and they are both good friends of ours. Rose is such a kind person and she's been a great inspiration to Margo this past year. She was ecstatic in her quiet way when she heard the news about Margo. We are all used to singing and playing together and we spent the next couple of hours singing for patients. We ended the evening by discussing our faith in God. They are not gung-ho religious types but are firm in their faith in a higher power. We discussed various signs from God and the proof is too powerful to ignore. The longer I stay here, the more I'm convinced in the grace of God.

Margo has had prayer groups praying for her from Winnipeg to Ashern, Cleveland to Prince George and our miracle is happening. You can't tell me all that praying hasn't had something to do with it. Being separated from Margo for the first time in many years has made me realize how much she means to me.

I truly believe I'm going through all this agony for one special reason. It is time to get down on my knees and beg for God's forgiveness. After Mum died, I ranted and cursed and shook my fists at the sky, asking God why He took her from us at such an early age. I believe now the answer is that He needed her for better things and it was time for her to go.

You've all heard the story about how I heard Mum's voice and felt her gentle touch a few days after her death. Mum must've been eager to go through and see her parents, her brother, and her dear son, Bob, but being typical Mum, she delayed crossing over in order to help her little boy get through that first terrible time after her death.

You would think after an experience like that I'd have turned to God and begged His forgiveness for ever doubting Him, but I turned my back on Him, continuing to hate Him, curse Him and deny His very existence. Through the years it only got worse.

When we are young and strong, we feel we are invincible and think we can do it all on our own. Over the last ten years, I've suffered from one health crisis after another. Sometimes, lying awake at night, I worried about what will happen to me after I die. I shunned church except for weddings and funerals where I listened to words that had no meaning for me. So many times God tried to reach me but always I denied His existence.

Wow, where did these words come from? I feel like I've just written a sermon. Actually, it is testament to my Lord, my Creator, and I never wrote the words, just held the pen. The hardest part was trying to stay far enough back so my tears wouldn't wet the paper.

I feel the tears I shed tonight have cleansed my body and soul and I have been re-born. If any of you don't believe me, please don't mock me. I hope Margo will come to believe in God. I love her so much; I want her beside me in eternity. I've already led her down countless rocky roads. How can she refuse to walk this gentle path with me? Love Dan"

It's hard to explain what happened to me that night. One minute I was almost freezing in that cold room as I tried to put my innermost feelings on paper. Something was building in me bringing me to the point where I couldn't go it alone anymore. I put my pen down and prayed to God, telling Him I needed Him in my life. The words were barely out of my mouth when I was immersed in a warm glow the likes of which I'd experienced only once before when I delivered the eulogy for Linnea. I'd felt His presence that day too but I'd been fearful of Him as though He might be going to punish me for turning away from Him. This time I was aware of His great love for me. I knew at once I was safe from all harm no matter what the future might bring. God had His strong arms wrapped around me holding me closely like a child as I cried my eyes out in relief and happiness. For long minutes I sat in silence trying to make sense of what had happened.

I was sitting there exhausted when a nurse's aid, Julietta, poked her head around the corner to check on me. I must've looked rather strange because she came close and asked if I was all right. I assured her I was very well. We started talking and when I told her about Margo and her latest good news, I broke down and the tears started to fall again. That sweet gal wrapped her arms around my head, rubbed my back, and assured me, "That be okay."

I quickly regained my composure and she invited me to join her for tea at the nurses' main desk. When I told her about my amazing experience, it was her turn to cry when she told me she was a Christian.

That morning I started a letter to Margie. It was the beginning of a new day, a new year and a new stage of my life. Neither Margo nor I knew what the future had in store for us but for the first time in my life, there was no fear in my heart.

"Dear Margie, Jan 1, 2004

I'm half-way through day 12 at Victoria Hospital. I'm looking out on the parking lot waiting for Tracy and Dylan to arrive. I've been getting many visitors. Jim's daughter, Dawn, works here and she visits me every day. Today she will be on this floor so I'll see more of her.

This horrible ordeal has turned into the most wonderful experience of my life. After faking it for many years, dutifully mouthing the Lord's Prayer at funerals and singing gospel songs at parties, last night I asked the Lord to accept my soul once more. A warm glow has enveloped me all day and my nurse tells me I have no fever. I've been a hypocrite all year praying for Margo to get better when I didn't believe in who I was praying to. Now I can put my faith in the Lord and lean on Him.

If I'm lucky enough to walk out of here on my own steam, I intend to start singing at the care homes again. I'll also see if the hospital wants me to go door to door to sing to any who want it. I have learned firsthand over the past month how slowly time passes when you are alone, scared and in pain. Maybe this will be a small way of serving my Lord and my fellow man.

I started today off as usual with a brisk 45 minute walk with a stop at the equipment room to weigh myself. I weigh 203 lbs, down from 232 lbs a month ago. Shower comes next, then breakfast, which is followed by an hour's entertaining.

I have quite a following here with more requests for singing every day. People hear me and wave me in to sing to their loved ones. It feels great to walk into a room where people are lying with their faces twisted in pain and a few minutes later I have them laughing. I can only imagine what it does to their stitches but it sure boosts their spirits. As much as it does for them, it is returned tenfold to me and my heart overflows with love for these dear people.

Yesterday morning, an old lady asked me to sing for her hubby of 51 years, Joe. He's a great old guy, a logger and farmer all his life. I tried to sing, "You're My Best Friend" for them. Halfway through the second verse I thought of Margo, choked up and stopped. I apologized and explained to them what she is going through.

I've had a visit from Dr. McFlicker who is taking over the investigation on my stomach. He'd already had more blood taken and says there are more tests coming. He said my condition is a real mystery. I don't want to be a mystery man. I just want to go home.

Jan 3, 2004 11:30 AM

After last night's storm blew past, we have a beautiful, cold, clear day with lots of sunshine. Now that I am somewhat well again, my heart longs for home.

I'm sure at first all the nurses considered me a somewhat queer but harmless old nut. Since then, they have seen me work miracles with the sick people here. Last night a patient heard me singing in the hallway and came dashing out to join me. Agnes Lucas has a fine singing voice and by the end of the second song, she was dancing. The nurses were amazed and told me later that Agnes has been depressed since she got here. I almost feel like I've missed my life's calling. Maybe I could've become a social worker.

A combination of Watson charm and good old-fashioned Irish blarney has worked its magic on the staff here and as I make my rounds twice a day, I'm greeted by radiant smiles and kind words from every direction. I always considered myself a good and decent man, but now that God has accepted me back into His fold, I will be a much better man. I feel totally blessed and truly at peace for the first time in my life.

Jan 4, 6 AM

Hi little sister. How wonderful that you phoned last night so I could tell you the story of my life's greatest experience. I know your tears were tears of happiness for me.

Yesterday, for the first time in almost three weeks I woke up longing to go home. I soon asked my day nurse to page Dr. Cowan so he could release me. I knew Hulda and Albert were on their way in to see me and I could catch a ride home with them. I waited all morning but the doctor never answered his calls. I had a nice visit with Hulda and Albert and told them about my renewed faith. Hulda is a true believer, even after losing her husband, Ronald, to a massive heart attack. She always has a serene look on her beautiful face and you can sense her inner strength and peace of mind.

Jan 5

I was up early and did my 40 minute walk. Julietta was falling asleep on her feet after working a double shift. She asked me to join her for coffee. She gave me two slices of toast with jam and peanut butter and it was so good. I won't have to eat hospital food this morning.

Dr. Cowan came in and we had a great talk. My newly found faith allowed me to speak firmly. I told him I had to have results soon or he could send me home. His eyes opened wide with respect and he started to talk man to man for the first time. He even asked me what is going on with Margo and listened attentively while I explained. He must be so used to looking at people as bodies to operate on but now he knows I have a mind and am not afraid to use it. I could never have managed this last week—another benefit of my new-found faith.

Jan 5, 9:22 PM

I'm going to close this letter with a few words about yesterday. Eunice visited with me and cried when she heard the wonderful news about Margo's tests. Tracy and Dylan arrived shortly after lunch, followed by Margo and our sisters, Ruth and Kathy. I thought we were having a great afternoon as I shared my story with the others but Tracy was becoming increasingly quiet and withdrawn. She left early and kissed me goodbye with a look in her eyes that I couldn't quite fathom.

Margo left with the sisters but phoned later from home to tell me Tracy was terribly upset with the change in me. She told Margo, "This isn't my Dad, but a total stranger." She was afraid I was going to flip out and have a nervous breakdown.

You've heard the term "high on God." That's me for the past five days. No wonder Tracy is afraid. I phoned her immediately and tried to allay her fears by telling her I am changed all right but it is all for the better. How much worse it would've been if the devil had entered my fragile soul instead of God. During this past terrible year, I could've snapped and gone back to drinking. The devil finds work for idle hands. I told Tracy to hang in there and trust me and she will see more amazing changes. I truly wish we'd taken our kids to church instead of copping out and letting them choose later. How can you believe in something you know nothing about? That is Margo's problem right now but we are going gently forward together. I'm not afraid of what the future holds for us anymore. Pray for us. Love, Dan"

It was strange to find myself so enthusiastic about God that I wanted to tell all who would listen about my amazing experience. Some of my friends must've thought I'd snapped under the pressure and they were probably right. If they didn't like what they were hearing, they were kind enough to keep it to themselves. My family members bore the brunt of my sudden passion for God, especially when the questions I asked them were much too personal.

"Dear Rose, 5:56 AM Jan 6/04

By now you will have heard about the amazing transformation that has come over me. I begged the Lord for forgiveness and He took me back into His arms and forgave me. I had cursed and vilified Him in the days, weeks and years after Mum's death. I've been reaching out to Him tentatively for the past ten years or so but it wasn't until He brought me to my knees before Him that I realized I wasn't strong enough to carry my burdens by myself. I have felt so scared and lonely and rejected lately, afraid I would die suddenly and go straight to Hell.

A few questions here, Rose and please believe me when I say I don't mean to offend you. If I do offend you, please forgive me and forget.

Rose, do you know if Dad ever came to an agreement with our Lord? After years of Mary reading scriptures to him and seeking out that one special preacher who could wave a magic wand and save his soul, poor Dad was tormented in his later years. I pray to God he is safe in Heaven with Mum.

How about you Rose, do you truly believe in God and Heaven? We've never discussed this together and I have no idea why the subject never came up during that long peaceful day on the river at Anama Bay last summer. Perhaps I was too intent on catching fish.

Yesterday I had a very extensive ultrasound and I know they found something. Halfway through the test, the technician called in her superior and they were whispering as they gazed into the screen as she continued probing my liver. I could hear, "See that? Is that an artery? No, that's a vein. Look right here. See that?" The movement of the probe in that area was very painful so I knew she was on the right spot.

Next she shut the machine off, went into the next room and paged Dr. McFlicker. I was sound asleep when she came back. If this had happened a week ago, I would've been lying in a foetal position with panic causing the pain in my belly to radiate to all parts of my body.

I got back to my room just ahead of cousin Dawn who had finished her shift. She took time to listen to my story and then said I should write a book about my experiences, starting with my childhood, right up to the present time. At first I laughed but it quickly became apparent that she was serious and maybe she was right. She said I could call the book Hitchhike to Heaven. I do feel I could write another book. We could get rid of the cattle and rent out the land. This would let me work on the book and spend more time with my dear Margo.

Dr. Cowan tells me the ultrasound showed a blockage in a big vein in my liver. A very strange case, he says, one that has baffled the doctors for a month. A CT scan is booked for 2 PM today. I'm so glad I'm getting action at last. After the tests, I'll probably be released and referred to a liver specialist. I'm actually going to be sad to leave here.

Wed, Jan 7, 2 AM

Well the big news is in. A mad cow that was found in the States has been traced back to the farm in Alberta where she was born. Of course, this could spell the end to the family cattle farm as we know it. Farmers might go bankrupt and have to leave their fourth generation farms. Busy farm towns might end up as ghost towns and frantic families will rush madly to the cities like lemmings falling to their deaths in the sea. My heart aches but I have no easy solution.

We've already chosen our escape route which should carry us safely into retirement. Last night, Margo asked Buddy Bergner, our auctioneer to phone me here and we discussed selling our entire herd in one large dispersal sale. He believes such a sale now would be disastrous. Buddy thinks if we split our herd into three groups according to age, he may be able to arrange private sales in areas where feed is no problem.

With the herd gone, we could easily sell our feed inventory and that would give us a year's income with no expense. The hay and pasture land could be leased out; we've had offers already, and we could buy that fifth-wheel camping trailer. We'd be free to spend all our time together for however much time our good Lord allots us.

8 AM, same day

I have time on my hands so want to relate an experience from a few days ago. This happened on my evening rounds. I happened to sit in a chair in front of room 409 and had barely started singing when a little man popped out. He started singing word for word with me. When the song was finished, we started talking. Ken is a dead ringer for Mickey Rooney but he was a frightful mess. His cherubic face was scraped from a bad fall. His left arm was in a cast, broken in three places from elbow to shoulder. He asked for Hank Snow and when I came through, he fairly danced with glee as he sang along. We spent half an hour getting to know each other before I told him goodbye. He was a puzzle to me, laughing one minute and then looking sad the next.

The next morning I did my rounds, feeling blessed from all the love flowing back at me. I left Ken until last because I felt something very important was about to happen. I found him sitting up in bed reading. He smiled at me with his angelic, blue eyes sparkling behind his wire-rimmed glasses.

We got right into singing and his feet were kicking the blankets right off the bed. Between songs he seemed terribly sad. I asked him what was troubling him and he said he wanted to go home but the hospital wouldn't release him unless he had someone to take care of him. I asked, "Don't you have any family?" His face fell even further and he said, "I used to but a few days ago I had a horrible fight with my wife, my son and his wife, and I don't think we'll ever make up again."

I put my guitar down and sat beside him on the bed. I took his hand in mine and could feel the emotions flowing between us. I told him the rift between him and his family was still fresh and now was the time to fix it. I told him the longer it festered, the harder it would be to make the first move. I said that three simple words, "I am sorry," would fix the hurt. He replied, "I know you are right, Dan. I've been doing some powerful thinking on it."

I asked him if he believed in God and he quickly replied, "Very much so." I spent the next hour telling him my life story and about my falling out with God and our miraculous reunion. He listened quietly as he lay back with his eyes closed. Once in awhile, he would question me or comment but for the most part, he was silent. When my testimony was done, he opened his eyes and said, "That was a very powerful story, Dan."

I told him Margo is going to start going to church with me and we want to start studying the Bible. A wise look appeared on that Mickey Rooney face when I asked him if I should start with the Old Testament. He said, "No, Dan, I suggest you start with The New Testament. There are too many discrepancies and distortions in the Old Testament for my liking." I burst out with, "That's what I've been saying for years. There are some things in there that I can't swallow."

We sat in an easy silence for several minutes until I commented that he seemed extremely knowledgeable about the Bible. He replied, "I should, Dan. I became an ordained United Church minister in 1957." You could've heard a pin drop in that room as I gazed at him. It was starting to make perfect sense. Once again, the good Lord had worked His mysterious ways in order to confirm my new found faith.

We talked some more, finding out bits and pieces about each other's lives. Ken looked completely relaxed and happy for the first time. Next morning as I made my rounds, I found him still looking cheerful. We sang a couple of songs in his room and I bade him good-bye. That afternoon I went to his room but found only rumpled bed clothes lying on an unmade bed and thought he must be down in physiotherapy.

After supper, I had a visit from Oliver Mulhull. We talked for awhile before setting off on my evening singing rounds. I introduced Oliver to my special friends, patients and nurses alike. Ken's room was still empty. A quick check at the desk revealed he'd been released while I was down having the ultrasound. I asked the clerk for his last name but she couldn't give it to me. So there it was—another case of two ships that passed in the night. It had not been for nothing. He'd given me some timely advice and I'd like to think he'd taken my advice and made peace with his family.

I rode with Oliver down the elevator to the main floor and walked with him to the front door. I was stopped in my tracks by a blast of frigid air which caused me to shiver as Oliver disappeared out the door through a cloud of billowing steam. It was time to get ready for bed but I felt strangely excited as if something big was about to happen. Turning away from the elevators, I headed east and then north along a long, narrow corridor. I passed a door marked Emergency Chapel and made a mental note to stop there on the way back.

I pushed through the double doors leading into Emergency Admitting. Here was a totally different scene from the night I was admitted. The waiting room was empty and there was no line-up of ambulances jockeying for position outside the door. I walked past the admitting desk and was accosted by the clerk who asked what I was looking for. I told her that sixteen days earlier I'd been registered there before being left in room # 9 where I went through six hours of my own private Hell here on earth. I asked her if I could spend a few minutes in there and she told me to take as much time as I needed.

I headed down that wide corridor which was strangely quiet now, no gurneys full of crying people lining the walls, no harried nurses dashing back and forth from patient to patient. Soon I found myself standing directly in front of room #9. The door was open and the lights were off. I felt compelled to step inside for a few moments to let the memories return. I stood in the dim light for several minutes, thinking about how my life had changed in such a short time. Looking back it was easy to see that I'd been just a tiny part of God's plan as Bill had told me.

Feeling utterly at peace with my world, I stopped at the chapel intending to spend a few quiet moments in conversation with God. I pushed the door open and entered the tiny room which held four wooden chairs. There were two messages on the wall printed on cloth and framed in wood. The words seemed intended especially for me.

One sign read, "We are at our best when we are good to others." After a lifetime spent helping people less fortunate and the last two incredible weeks spent here in the hospital, this had to be more than mere coincidence. Since I found the Lord, I've been told countless times by nurses and staff that I have a special glow radiating from me as I speak with sad, pain-filled patients.

The other sign touched my heart as I read, "Your inner beauty shines forth from within." Overcome with emotion, I sank to my knees and the flood gates opened. I cried until I was exhausted, praising the Lord and thanking Him for taking me back. Thirty-seven years of rage, guilt and self-denial were washed away in that moment. Never again will I have to face difficult situations while standing alone with fear in my heart. I am safe in the arms of the Lord, my Creator. I'm sure you can all feel what a tremendous experience this has been for me. I feel totally drained but tremendously happy."

January 8 was clear and cold. After lunch, I settled down for an afternoon nap. I'd just fallen into a deep sleep when Ruth arrived with our cousin, Bill. What a pleasant surprise! I spent an hour sharing my experience with him. He cried when I cried and was totally happy for me. He'd been praying for my soul for years but knew better than to push me. I told him how we'd resented people trying to push religion on us and he said he can't stand that either.

I grabbed my guitar and the three of us went out into the hall to do some serious entertaining. Within minutes, that long corridor was jammed with nurses, aids, cleaners, orderlies, visitors and patients. In room 420, we visited Enid McRuer, a delightful lady. I'd told her that I'd written a book and she said she'd written one too. We had agreed to exchange books and her husband, Bob, had brought one in for me. I was astounded to learn she'd been a doctor all her life. You look at these frail, old people and would never guess at the brilliance within. A quick song for her, a hug and we were off to the north end of the corridor. The crowd followed us and stood clapping as we belted out two gospel tunes. The hall was bedlam now with the nurses changing shifts. The occasional doctor on his rounds looked totally confused at the scene. We made our way back to the south end with my friends reaching out to touch me. I was still in my street clothes so they thought I was going home.

Ruth and Bill dashed away and I'd barely sat down when Einar and Elsie arrived. A few minutes later, Dr. Cowan walked in. He said the CT scan showed a blood clot blocking a large vein leading into the liver. He'd found a liver specialist who would take my case, a Dr. Lipschitz, at the Health Science Center. We would be hearing from him soon. He asked if I could go home in the meantime and I was more than ready.

Einar offered to drive me home after they made their weekly visit to Elsie's sick sister in Deer Lodge hospital. I made a quick run from end to end saying good-bye to friends and staff.

Elsie's love worked its magic with her sick sister and after supper at their place in Woodlands, Steven met us half way to save Einar from driving late into the night. At home, Margo was fast asleep but she woke when I turned on the light. It was wonderful to see, touch, and once again hold my little sweetie. I was chilled and feverish for several hours as we lay there and talked, until the Tylenol set things straight. We talked about selling the cattle at once and were in agreement that it was the right thing to do.

At breakfast next morning, Buddy called expecting to convince Margo to hang onto the cattle. He was ready to call our neighbours to help look after our cows until I was well. He said he'd even come for a week at a time to calve our cows. I was deeply touched at his offer. We had so many friends phoning non-stop with offers of help. We were truly blessed and looking ahead with hope in our hearts.

EIGHT
From Elation to Despair

It felt so good to wake up in my own bed with my wife beside me. We were a sorry looking pair but we were together and that was all that mattered. Margo had dark rings under her eyes and her face was a sickly shade of grey. The skin on her feet and hands was peeling off in sheets and with the pain in her hip, it was all she could do to hobble around with the help of her cane. I looked more like a walking skeleton than a real live man but my appetite was improving and my strength was returning. It did seem as though we had turned a corner in our lives and we felt that things were only going to get better.

On January 10, we attended a dance in the Eriksdale Recreation Center. It was such a positive experience that we were pleased we'd made the effort to go. From the moment we stepped in the door we were enveloped by the love of our friends and there was never a dull moment. People kept arriving at our table with hugs and words of encouragement until we'd been greeted by almost everyone there. We stayed until after midnight.

The following morning we attended Sunday morning service at the Evangelical Free Church in Eriksdale where my cousin Bill Watson was the preacher. Bill delivered his sermon in a very relaxed manner with lots of love flowing in the room. It was a small congregation but the building was small too and it all fit well together. For the first time in years I felt that the service was directed straight at my heart though Bill assured me later he hadn't intended that.

At one point we were asked to share things we were happy about. I told them I was happy to have my faith in God and so thankful to have Margo still at my side and feeling better. It was so laid back and cheerful with lots of laughter that even Margo relaxed and enjoyed herself, which is how it's meant to be.

That night we dragged out the Sequence board for the first time since I'd been ill. Margo was perking up after her last chemo session and this time she'd been given an extra week to recuperate between treatments. We couldn't stop smiling at each other as we contemplated a bright future.

Within a week I was in Winnipeg for my appointment with the liver specialist. Dr. Lipschitz said my condition was rare but not unheard of. He prescribed blood thinners and booked an MRI to locate where the blood clot came from. He was concerned that it came from a tumour that didn't show up on the CT scan. He also mentioned possible damage to the lower bowels which may have been caused by loss of blood flow through the liver.

Three weeks previously I would've been filled with terror upon hearing his words but now that I had God taking charge of my affairs I was completely relaxed as I laughed and joked with this kind-hearted man. It felt so good to not have to worry about a thing as I put myself completely in God's hands. Now that my faith had returned, I wanted so badly for Margo to share it with me.

"January 15, 2004

The good Lord continues to send me reminders to keep my faith strong. Margo and I were talking about my newly-found faith and about her complete lack of faith in God or in eternal life hereafter. I mentioned that I want to buy a New Testament so we can read a bit together in the evenings. She disappeared into our bedroom and came back with a brand new bible which our friend Lois Wilson from Powell River had sent her. I was overjoyed to see that it's written in contemporary style which even a dummy like me can understand. I must confess I never enjoyed reading the bible with thee and thou and begat in it and could never imagine myself quoting scriptures from it. A quick glance through this one has me anxious to get started on it."

Buddy Bergner had found a buyer for most of our cows. On January 17, he brought Jim Quintaine, a cattle buyer from the Brandon area, to look at them. Jim brought Doug and Claire De'Athe and their daughter Signi who were ranching south of Carberry, Manitoba. They planned to lease our cows after Jim bought them. They walked through the herd and were impressed with how quiet the cows were and mentioned the cattle were in very good condition. I was hoping Jim would take the whole herd but due to the BSE scare, he wasn't interested in buying anything over six years of age. We'd been keeping our herd very current and Margo brought out records to show we had 110 pregnant cows under that age.

The days passed and both Margo and I were gaining strength. I was gradually taking over the chores even though Steven and Ruth were still coming out every day to feed the cattle. On Jan 19, I started a letter to our daughter which expressed the optimism we were feeling.

"Things are starting to settle down a bit but we still get enough action to keep us busy. Mum is getting stronger again as her blood count continues to rise. She gets four weeks at home between these two treatments so she still has about two weeks left to enjoy better health. I pray for the day that her tests come back clear and we can get on with our new life.

My own health continues to improve. It seems like the blood clot must be dissolving, allowing better blood flow to my liver and lower bowels. Yesterday for the first time in five weeks I was able to eat a good meal without a whole lot of discomfort later.

Buddy phoned to tell us the cattle buyer had made an offer on our young cows. He would pay the freight, plus the cost of the pregnancy test and had asked Buddy to supervise the sorting of the cows. Bud also offered to go with me to Brandon to observe the pregnancy test.

Mum and I sat down with Ruth and Steven and after talking it over from all angles we decided to accept the offer. It would lessen my work load to the point where Steven could go back to work. I could calve out the older cows and by getting rid of two thirds of the herd, we'd be free to sell most of the hay we have on hand. That would make up for the income we'll lose in this year's calf crop."

Not all of our news was good.

"Mum was called in to Dr. Steyn's office to discuss the results of the latest CT scan and it was unsettling news. There are more spots on the liver now even though the blood work showed the main tumours were down by seventy per cent. We don't know what to make of this but will have to wait for the results of the next CT scan in about a month. This is about par for the course with this rotten disease. Just when you think you are winning, it deals you another low blow."

That afternoon we sorted off and ID tagged 109 head of prime breeding stock. By selling the young cows and keeping the older ones, we were going against everything we had practiced for years. There was no choice in the matter because there was no demand for older cows.

A few days later I got a call from Claire De'Athe. She told me the cows had arrived and all of them were in fine shape. She was very impressed that the cows were extremely quiet. She and her family felt like they'd received a second chance to remain in farming. They had land and buildings but no money to buy more cattle and this deal had come along in answer to their prayers. They are Christians and have a deep faith in the Lord and it was such a pleasure to talk to her. I told her we'd been praying to find a buyer for our cows and God had answered our prayers as well as theirs.

On the third day of February Margo had another chemo treatment. To our surprise her doctor seemed elated at her progress. Her last CT scan looked fine to him and the tumours had shrunk some more. Once again, we found ourselves riding that emotional roller coaster as our spirits rose with this latest good news.

The chemo made Margo very sick but she perked up a bit when Tracy brought Dylan out to spend a week with us. Margo may have been too sick to move around but she could still cuddle The Boy as we had nicknamed him. That little guy did a lot to lift our spirits.

The last week of February saw Margo back in the Cancer Clinic. I stayed with Margo during her chemo treatment so I'd be there to help her through her nausea. Once we had her settled in her familiar room on the third floor of the Health Science Center, I headed home to look after chores before it got dark. It snowed all that night and the highways were bad the next morning so I was late picking Margo up. She was worried about me because she'd heard reports of highways blocked with heavy drifts.

For several weeks I'd been meeting cousin Bill at the hospital every Wednesday evening. I played the guitar as we sang for anyone who was in the mood for some gospel music. After we made the rounds of the hospital, we'd walk through the doors into the Personal Care Home and serenade the old people there. A lot of them were old friends whom I'd known since I was a child. It was such a pleasure to see how they brightened up when they saw us arrive. One little sweetheart, Annie Pottinger, never failed to tell us how much joy we were bringing them. The thought crossed my mind that they brought just as much love and joy to Bill and me as we gave to them.

On March 8, Margo came out to the barnyard and watched the gate for me while I moved several cow-calf pairs out of the maternity pen and sent them on their way to the feedlot out in the bush. Margo had always loved being around our cattle and she was badly missing watching the arrival of those beautiful little babies. No matter; we felt certain that within a year she'd be right back out there with me as we welcomed in a new crop of calves.

Later we headed toward the city for my MRI scan. Margo could've stayed home but she insisted on coming along to lend moral support. We were there half an hour early but they took me right in. It wasn't long before they slid me into a very narrow tunnel with my arms extended straight out over my head. There was barely enough room for my shoulders and my elbows were jammed inside the tunnel. I suffer from a bit of claustrophobia and I felt panic rising but managed to stay calm for the 45 minutes it took to complete the test. Margo was waiting patiently for me and we walked down those familiar corridors and made our escape.

"Last night, Margo was really sick as she usually is at this stage between treatments and she was on her way to bed shortly after seven. I talked her into playing Sequence and she promptly beat me three games straight. That Sequence board has been a life saver for us during this stressful time.

No matter how sick she is, she always rises to the challenge. Once the cards are dealt we show each other no mercy and none is expected."

On March 11, we were back in the city for a follow-up with my liver specialist. He told me I was in pretty good shape with no sign of a tumour on my pancreas or elsewhere that might've caused the blood clot in my liver. New veins had formed around the clot and blood flow had been restored. He was arranging to have me see a blood specialist at St. Boniface Hospital to see why my blood had clotted the way it did. If they found that everything was normal, I could be off the blood thinner in about six months.

March 16 brought the start of another three day marathon for Margo when I drove her to the city through a violent late winter storm. We'd planned on bringing her home after the special blood tests but didn't want to risk not being able to make it back to the city the next day for the chemo. I dropped her at Kay's house and headed back. It had snowed six inches while I was gone and my little car was dragging all the way.

"March 17, 2004

Margo phoned to say her blood test was good and she'd be having the chemo treatment this morning as planned. Poor little gal is getting tired of those darn treatments that make her sick for two out of every three weeks. I don't know how she finds the courage to go back time after time when she knows full well that she will be extremely sick. I'm praying she'll soon be in remission and will get a rest from this nightmare.

March 19, 6 AM

Margo was completely wound up by the time we got home and I couldn't get her to sit down and relax. I think the drugs do something to her system that makes her really hyper for a few days.

The news she got from her doctor wasn't exactly what we wanted to hear. The cancer markers have stabilized but they still aren't where he'd like to see them. He wanted some input from us regarding when we should quit the treatments. She told him she wanted to leave that decision up to him. He is the expert and we have to trust him to make the right decisions. We don't want to see her quit the treatments too soon as long as there is a chance they are still doing some good. She is booked for a CT scan and another treatment in April and I'm going to make darn sure I'm there with her when she sees the doctor next time. She has been going it alone way too much since I got sick in December.

The receptionist from the doctor's office told Margo they may take a look at the scan and decide the next day to stop treatments. We have mixed feelings about that. We want to see an end to these awful treatments but we don't want to stop too soon only to have the cancer come back. After all she's been through, we want a remission that will last as long as possible. Margo has been feeling very down, not knowing when the treatments will end and what the future will bring after that. I feel so helpless and wish there was something I could do. If someone was trying to harm her I would gladly give up my life defending her but how can you fight something you can't even see?"

Early April brought news that Margo's cancer markers had come down again and the chemo treatment went on as planned. We'd been fretting because we knew if the markers didn't come down, we'd be expected to make a decision. Ruth took Margo to the city for the treatment and stayed overnight there. They got home about noon the next day. Margo was feeling fine but very hyper. She never got sick this time and had supper as well as a bed time snack. She never ceased to amaze me with the way she coped with whatever was thrown her way and still came out smiling.

That weekend we threw caution to the wind and left the cows to look after themselves while we went to a dance in town. Margo was feeling quite well and we danced several dances together. She even wanted to try a polka but I talked her out of it for fear of aggravating her sore hip. Time flew and before we knew it the dance was ending.

"April 14, 2004

Mum has been very sick since the last treatment and I'm not too sure she doesn't have a touch of stomach flu. She's worn down mentally from the never-ending torture and we both look forward to the magic day when the doctor says, "No more treatments." She's been very strong through all of this, much braver than I'd have been, I'm sure."

Near the end of April Margo came with me to a fundraising variety show in Lundar on a Sunday afternoon. The show was based on the Grand Ole Opry and I played the unlikely part of Grandpa Jones. I'd spent three weeks learning a few chords on the banjo and got lots of laughs from the large crowd there.

"It was a fun afternoon but a very long day, especially for Margo. Because she was tired, we decided it would be easier to go to Chicken Chef for supper. When Margo stepped out of the car, a strong gust of wind whipped her hat off her bald head and sent it tumbling across the parking lot. I had to chase the darn thing almost all the way to the goose statue before I finally stopped it by stepping on it. I told Margo I could've done better if I hadn't been distracted by the sight of a bald headed, old guy running along behind me screaming, "Get it! Get it!" We do have our laughs together even during the worst of times and that keeps us going. Margo was still laughing as we sat down inside the restaurant and once again, I marveled at her amazing courage."

Two days later we drove to the city for another blood test and stopped on the way home for lunch with our friends, George and Jean Griffin. They are very fine people who we met several years earlier. We often saw them at dances and they always asked us to stop in and see them.

The following morning we headed back to the Cancer Clinic and though we never spoke about it, I know we were both dreading what the doctor might tell us.

"We waited for close to an hour for the doctor to arrive. He took plenty of time with us and the news wasn't what we wanted to hear. Margo's tumour markings have come down again from 40 to 34, but she has to continue having treatments if she wants to stay alive. I asked him what the chances were that she would enter remission. He told us gently that the chances of that happening are very remote. No matter how low the tumour markings become, there will always be some cancer cells lurking there waiting to start growing as soon as the treatments are stopped. He also confirmed what we'd already suspected, that the longer she stays on chemo, the worse her side effects will get.

He says she can quit the treatments and see what happens or else she can go off them for awhile to give her body a chance to get back to normal. We compromised on the deal when he offered to extend the period between treatments from three to four weeks. That way she can have two weeks where she will feel somewhat better and she won't need any blood tests during those periods either.

The doctor asked how I'm doing with my liver trouble. I told him I need to see a blood specialist but they don't have time to take me in for something as trivial as this. He looked at me over his glasses and told me he would see that I got in shortly.

We were barely seated in the waiting room when Dr. Brandes' nurse appeared. She'd already contacted a blood specialist and I have an appointment on May 7. I really appreciate the doctor helping me out like this and we realize now he isn't the heartless individual we thought he was when we first met him. These poor doctors try to keep a safe distance between themselves and their patients so they don't get emotionally involved but it doesn't always work out that way.

As soon as they took Margo into the chemo treatment room, I kissed her goodbye. I then went to see the The Passion of The Christ. The big rush is over and the theatre was almost empty. Twenty minutes into the movie, I was weeping like a baby and never really stopped for the rest of the show.

April 30, 8:10 AM

We were barely back home yesterday when John Webb phoned. He's interested in doing a follow-up on the BSE situation one year after the fact. I told him we've sold most of our cows and are thinking more and more about retiring."

More and more I found myself believing I am just a tiny part of the Master's plan. On May 3, I wrote,

"We went to church yesterday morning and Bill preached a good sermon. I always get some little message that seems as though it's meant just for me. This time the message came through loud and clear. Love is patient and kind. How true that is! I love Margo with all my heart but with all we've been going through lately, I sometimes lose my patience because of something she says or does because she is feeling so bad and we end up squabbling for no real reason. I came away from church with a firm resolve in my heart to be much more patient and kind to her from now on. She doesn't need any unnecessary hassle from me and most certainly can use all the love and patience I can give her."

I believe if there was any defining moment in our relationship, it was when that simple message entered my brain and became implanted there. From that moment, it became so easy to overlook little things that might've bothered me. From that day, our love blossomed and grew until we became so close it was as if we were one being and so deeply in love, it was almost hard to bear.

May 4 was my sixty-second birthday. Steven and Linda dropped in to tell us the wonderful news that they were expecting another baby. We'd been feeling pretty downhearted and this good news did a lot to cheer us up.

On May 7, Margo came along for moral support when I visited a blood specialist at St. Boniface Hospital. I was given a brief examination and several blood samples were taken. We were told it would take eight weeks for the results of some of the tests.

On May 15, we took in the old-time dance in Eriksdale. Margo was feeling very weak but she perked up after we got there. It turned out to be a great dance and we both danced quite a bit. We even danced both sets of a square dance with our good friend, Bill Beauchemin, doing the calling. Looking at Margo's happy eyes and smiling face it was hard to imagine there was anything wrong with her.

John Webb showed up to do some more taping for CBC on May 19. I'd tried to tell him that I didn't have anything to say that hadn't been said at least a hundred times already but he was very persuasive. We ended up doing an interview out on our deck. From there he drove up to Ashern to tape part of the cattle sale and the resulting tape aired on May 20, the anniversary of the BSE crisis.

On May 25, we were in the city for Margo's blood tests at the Cancer Clinic and left for home after having lunch with Tracy and The Boy. Early next morning we were in the city again, dreading the thought of another session of chemotherapy.

"The doctor thinks Margo needs to take some time off from the chemo to let her body rest. Her feet and legs are swelling badly and she has other side effects that are making her life miserable. What a cruel disease this is and the treatments are just as bad. He gave her another four weeks off and then will do more blood tests to see where things are heading. Margo was so relieved to skip the treatment, she almost cried."

On June 8 we made another trip to the city—this time trying to kill three birds with one stone. We made a quick stop at the Health Science Center to see my liver specialist before zipping over to the airport to drop off Ruth who was heading out to the west coast. We also met Bertha and Dave's plane. They were coming from the west coast to spend some time with us. The next few days we saw an amazing change in Margo so we knew her big sister was the best medicine she could've had.

On June 18, we drove our guests back to the airport after a ten day visit. The time had passed much too quickly and we were sorry to see them go. Margo was feeling stronger every day and her voice, which had been sounding old and scratchy, had changed until she sounded like a young girl again.

On June 22 we made another quick trip to the city where Margo had blood tests and a CT scan. The next morning we were back in the city again to have a conference with Margo's doctor.

"The CT scan showed the tumours on her liver were stable but the markers in her blood were coming up again. As he told us, cancer doesn't take a rest when we do; it just keeps on coming back. The doctor suggested we cancel one more chemo treatment to let Margo regain strength and then he wants to put her on chemo pills. If the chemo pills can keep her cancer under control she'll be able to live a better quality of life with fewer side effects. This was great news for Margo who had been absolutely dreading going back on chemotherapy. We realize now why some people on chemotherapy finally come to the decision that enough is enough. I have reservations about postponing the treatment because if the tumours on her liver expand or increase in number she will be even sicker than she is now.

That evening I dashed off to meet Bill at the hospital. We went from room to room singing to anyone who wanted to be entertained. In the palliative care room we found Toots Paulson, the longtime wife of Dr. Paulson, the man who delivered me, some of my siblings and also my son, Steven. I shared with her the story about the time her husband was checking out the wound in my butt after it got impaled by the grapple tooth on our front end loader. When I told Toots that I don't know how Doc managed to find four inches of meat on my skinny butt, she laughed until she got tears in her eyes.

Toots asked me about Margo's health and I told her Margo just has to live with the disease as long as possible. Toots looked me straight in the eye and told me she is beginning her own final battle with cancer. I believe her because once you are in the palliative care room, there is usually only one way out.

Elaine Henrotte, one of my first teachers, was in the room visiting with Toots and she sang along to a couple of beautiful hymns. It felt good knowing I could help bring some happiness to the widow of a man who was such a major influence in this part of the country."

On June 25, we attended a funeral in Lundar for another friend, Cliff Olsen from Oak Point—another cancer victim. Cliff and his wife, Dorothy, had visited me in the Victoria Hospital when I was down and out and it was good to be able to pay my last respects to Cliff. It felt as though the whole world was dying from cancer and it seemed like almost every day we heard about another friend or acquaintance that had been diagnosed with this horrible disease.

That afternoon as I worked at getting my field sprayer ready to spray herbicide on our barley fields, I wore rubber gloves on my hands and a mask on my face. It was easy to remember the time when all farmers worked bare-handed as we cleaned plugged sprayer nozzles before lifting them to our lips to blow through them to clear the obstruction. We didn't know enough to be cautious and it was little wonder that at least five of my neighbours in the fourteen mile stretch between our farm and town had been diagnosed with prostate cancer. All had been successfully treated but the fact there were so many of them was cause for alarm.

Why had Margo come down with cancer when I, who had literally washed my hands in weed spray for years, was still walking around healthy? I thought about all the bottles of hair spray and deodorant and other fancy stuff that filled the shelves in our house. Margo was a sucker for anything that came in an aerosol can whether it was fly spray or perfume. For years we carried on a running argument because she insisted on using a type of fly spray that automatically sent a shot of toxic vapour across the kitchen every fifteen minutes or so. It was extremely deadly alright and every morning we'd sweep up all the tiny carcasses of flies that had been unlucky enough to move into our house. My argument was that if the darn stuff was strong enough to kill flies overnight it would surely kill us too, given enough time. Her response was logical enough. The product had been approved for use in restaurant kitchens so it should be safe enough in our house.

I couldn't help thinking that even though she'd been strong enough to win each battle in our war of words, she was now losing the battle for her life and she was paying the supreme price.

On July 2, we drove through a downpour of heavy rain to see Dr. Steyn. Margo had been having lots of back pain and spasms around her lung area. The doctor brought out the records that had been sent from Dr. Brandes' office. The CT scan showed that Margo had lesions on her lower spine but it wasn't clear if the lesions were from cancer or arthritis. Dr. Steyn sent her to the local hospital to have a bunch of x-rays taken.

We spent that afternoon playing Sequence with Hulda and Albert as we watched the rain pour down. Times spent with wonderful friends made life bearable during those dark days when it was hard to work up a good smile. Our friends kept in close touch by phone calls and sometimes surprise visits.

"July 6, 2004

I was working like a beaver hauling logs out of the brush pile when I saw our truck pull up with Margo driving. There was another couple in the cab with her, our friends Elmer and Linda Nickel from Steep Rock. Elmer hopped into the tractor cab and rode home with me while the women went on ahead. We hadn't had a good visit with them for years and we made the most of it. Nickels are experiencing their own health crisis since Linda was diagnosed with MS last year. It is the slow moving kind and the doctors feel that with proper medication she'll live a normal life for many years yet."

Hay-making got off to a late start because of all the rain. Margo was feeling stronger and cooking all the meals again. It felt so good to come in from the fields and find her waiting for me with a hot meal on the table and a smile on her face.

On July 19, Margo felt well enough to drive herself to the city for blood tests. She stayed overnight with Kay and Alf so she'd be there to see the doctor in the morning.

"Margo gave me the bad news that her tumour markers were coming up again, as had been expected. She lost another six pounds this past month mainly because she has no appetite. The doctor wants to start her on chemo pills which are going to cost us $1,340.00 per month but if they work we'll spend every last penny we have. We're already spending $200 per month on some useless herbal pills.

The chemo pills won't be as hard on her as the chemotherapy but they'll give her some bad side effects just like the intravenous chemicals did. Her hair has been growing back and apparently it will keep on growing and she is so thankful for that. Isn't it wonderful that in the midst of all this pain she can still find herself thankful for small mercies?"

A few days later we were in the city for an appointment with my liver specialist. My recent tests showed no signs of cancer and he wanted me to stop taking the blood thinners. As we drove home I had very mixed emotions running through my mind. On one hand I was thankful my tests had come back clear but I also felt guilty because Margo was getting worse instead of better. Why should I be so lucky while at the same time she'd been dealt a death sentence? It wasn't even something that I felt comfortable talking to her about, at least not at that stage. I knew it was not for me to question such things but over and over again I asked, "*Why, God, why?*"

We were home by supper time and after a quick bite to eat, I headed straight out to our Chippewa quarter to bale hay. The heavy swaths of tame hay lay glistening in the sun in the old lake bed where bulrushes once grew in several feet of water. The swamp had been completely drained many years earlier and had become some of our most valuable farm land. Margo, Ruth and I had worked countless hours in this area in the summer of 1996 picking rocks from the peaty soil. Over the course of that long summer, working in our spare time, we picked 135 loads of stones until we had the whole field looking like a garden. Now as I rode over this same fine field, I recalled the hours of hard work, sore backs and aching muscles. Tears ran down my cheeks when I thought about how much I loved this land and my dear wife who had worked so hard alongside me all these years. There was no time to stop the tractor; I wiped the tears away and kept driving. The hay was dry and that black storm cloud was getting closer by the minute. I worked until midnight until the last swath was in the bale and then drove home to crawl gently into bed beside Margo, the love of my life.

On July 26, Margo was feeling well enough to take the car to our dealership in the city to have some service work done. Once the car was ready she drove out the Trans Canada Highway to pick up ten baskets of strawberries at Portage La Prairie. At four o'clock, I came in from the field for an early supper and found Margo in the kitchen hulling berries for the freezer. By the time I came in from baling at dark she'd processed the whole works. It amazed me to see how strong she was becoming, now that the chemicals were leaving her body.

The next day a phone call from Dr. Brandes' office informed us that the x-rays taken in Eriksdale Hospital showed that Margo had a fractured rib. She hadn't taken a fall so this was rather puzzling. We prayed the cancer hadn't entered her bones.

Everyday farm life went on as normal but there was always that dark cloud hanging over my head, making it impossible to sleep all night even when I was exhausted.

"August 9, 3:17 AM

Here I go again, wide awake in the middle of the night. My mind is brimming with thoughts of what tomorrow might bring when we visit Margo's oncologist. We go to the city today for her blood test and then back again tomorrow to see the doctor.

By the time we left for church yesterday morning it had started to rain but we got ahead of the rain cloud as we drove. We arrived just ahead of the storm and it rained all the time the service was going on. God was looking after our needs even as we sat and sang while worshiping Him. It has been hot and dry for a month and the land was drying out badly.

Every Sunday we have a prayer request and praise session which I really enjoy even though I seldom take part in it. Yesterday I listened and prayed as usual as several people spoke up with special concerns in their life. Then I spoke up with two requests of my own. I asked the congregation to pray for a good report from Margo's doctor tomorrow and also that she will come to know and love God as I do. Several people stood up and prayed for Margo to recover or to continue to live and function in a normal sort of way. There was so much love flowing in that room and I could feel God's presence all around us. I held Margo's hand and wiped tears from my face with my other hand. I've felt such peace in my life since I came back to God and never again will I fear death. It is a strange, but wonderful, feeling for someone who has lived in fear of death all his life as I've done."

On the way home from church we checked the cows in the far pasture and found one limping with foot rot. We'd have to bring the whole herd into the corral to give antibiotics to the lame cow. We drove home and had lunch before heading back to the other farm with the truck. We'd worked as a team for so many years and that afternoon was no exception.

"We closed the gates around Ruth's house yard and then drove out in the pasture and called the herd over to the truck. Margo drove ahead calling the cows while I followed along bringing up the stragglers in the rear. The cows followed her for the full half mile and straight on into the barnyard. It was all I could do to run fast enough to close the gate behind them.

I got the corral gates ready and Margo helped me drive the cows in. She was following along behind them through wet grass where she got fresh cow manure on her pants and her jacket. She totally loved every minute of it. She sure got her blood up and wanted to do more and more, saying things like, "Do you want me to run over and open the pasture gate so they can go straight out when you open the corral gate?" I had to remind her that her arm bone might break if she used it to open that tight wire gate. It felt so great to have her working with me. It was just like old times when we did all these things together. In those days we completely took our good health for granted.

Back home, I picked fresh lettuce, radishes and green onions to make a salad for supper. Margo cooked steaks while I made the salad and we ate like kings. It was a perfect ending to a perfect day."

On August 9, Margo had her blood tested at the Cancer Clinic. The rain continued so we weren't worried about losing precious haying time. The following afternoon found us back for an appointment with Margo's doctor. We had mentally prepared ourselves for the worst.

"The doctor took time explaining things. We'd been desperately hoping for good news about the results of the new chemo pills Margo has been taking but we ended up getting more bad news than we'd bargained for. Her tumour markers are still going up but he said that it often takes at least three sessions of those pills before the markers start coming down again. The real shocker came when I asked him why she should have a fractured rib when she hadn't done anything to hurt it. He laid a bomb shell on us when he said the cancer is in her bones.

The real danger here is that as the cancer eats holes in her bones, she could break anything, like her hip bones for instance, which would almost surely be the end of her. The doctor is setting up a treatment which will be done at the hospital out here under Dr. Steyn's direction. It's a three hour intravenous drip which goes into the holes in her bones and bonds with the calcium to make her bones strong again. He also sent her to x-ray where they filmed her bones from head to toe to look for damaged spots. He may prescribe radiation on the hot spots and here we go again. How much more of this torture can she take before she says, "Enough is enough"? Those radiation treatments are usually five days a week for five weeks which means we would be apart for most of that time when all we want is to stay as close together as we possibly can.

We are both feeling bitter and sort of in the "why us" mode but I know I will never lose faith in God again, no matter what happens. My biggest worry is that Margo won't find her faith in God before she passes on and we won't get to walk together in Heaven. I don't push her at all because I know she hates to be pushed as much as I do. I pray with all my heart that by going to church with me, something will twig in her heart and she will become a believer. I have such a different feeling about death now. If someone came to tell me that I was going to die tonight, I would spend the day in a state of high excitement knowing I was going to meet my loved ones soon. I wonder what Mum looks like now. Is she old and grey like she was when she died or is she young and beautiful like a young girl? I wonder if Dad and Bob made it to Heaven. I pray they did and that I will see them again some day.

August 12, 7 AM

Wow! What a day yesterday turned out to be. After spending all winter weeping in private to keep up a brave face for Margo, I finally lost my cool and cried like I couldn't stop. This time it was Margo holding me close and comforting me. I'd spent some time on the phone talking to friends and their love and concern got through my skin which is thin at the best of times. Well, at least Margo knows that I do love her and she appreciates that. I think I may have learned a valuable lesson here. If you hold back your emotions in an attempt to remain strong, your loved one may get the feeling that you don't care all that much. For the rest of the day we couldn't keep our hands off each other and I feel we are even closer than we've ever been. It reminds me of the words in an old song, 'I love you more than yesterday but less than I will tomorrow.'"

That afternoon I had a call from Tiny Lamoureux, who excitedly told me he'd been in touch with a doctor in the city who felt he could help cure Margo's cancer. This man claimed to be able to cure an astonishing variety of diseases, including cancer of all kinds. Tiny was planning on taking his wife, Arlene, to meet this man in hopes that he could help her with various health problems and he wanted to know if we'd like to go along. This all sounded too good to be true but after talking it over with Margo, we decided to at least listen to what the man had to say. We'd been told by her own doctor that there was no more hope for her so why not try some alternative medicine?

"We were face to face with this man who might be our last hope. He went into his well rehearsed spiel trying to make believers of us. I listened carefully to everything he said but in the back of my mind I was thinking, "This man is either a nut case or a scam artist."

He went on and on until my head was reeling with facts and figures and I finally had to remind him we had to be back in Eriksdale that afternoon to see our family doctor. I asked for an estimate of what it was going to cost and he came up with a figure of between $300 and $400 per month for the different herbal and vitamin pills that would be required. He told us she would be free of cancer within a year. When Margo heard those words, she turned to me with hope shining in her dark eyes and a brilliant smile on her lips. In spite of my doubts, hope was rising in my heart too. This man seemed so positive, it was impossible not to feel the optimism. He would email the information about Margo's condition to a clinic in the States and a program would be worked out regarding her treatment.

The crowning touch was when he took us into another room where Margo placed her hand on a Bio Scan Machine which would monitor her condition each month. This fancy machine is supposed to be able to pinpoint any health problem but I noticed that he kept us talking until we'd told him everything we knew about her condition. Not until then did he consent to give her the test. The cost of the initial bio-scan would be $125 and $25 per month after that. He'll download the info and send it to the States and by this evening he'll get Margo's report back. We'll see him next Monday and she'll start taking the capsules and a special diet to follow.

So there you go. We might be risking a few bucks but the chemo pills cost much more. Why not give this a shot? At the very least, it will give us hope again and that is so important to us right now. This past year and a half has been a constant state of seeing our hopes dashed in a steady downward spiral. This might turn out the same way but we'll be able to say we tried. I couldn't live with myself if I let her suffer and die without trying every possible avenue."

That week brought a visit from Margie and Larry who had driven all the way from Sardis, BC, to spend some time with us. We had Tracy and The Boy out as well so Margo was in her glory. With new hope in our hearts we attended the dance in town on Saturday night and stayed until they closed the doors.

Monday morning we were in the city to see what our new doctor had to say. All weekend we'd been thinking about this new treatment and had pretty well made up our minds to go for it even though both of us felt a little doubtful.

"The doctor had a serious look on his face as he told us the case was much worse than he'd thought. The report had come back from the States and it showed that Margo has three types of cancer, in her breast, liver and bones. I felt like saying "Oh, really? That is exactly the same information we gave you last week," but managed to keep my mouth shut until I heard what he had to say.

Before he got the report back, he'd drawn up his own plan about the supplements to put her on and the total cost came to exactly $396.82 per month. Surprise, surprise again! He'd quoted us a ballpark figure of between three and four hundred per month and this figure was conveniently just a few dollars less than that.

He said the report advised having her go on a liquid supplement which is in a new category of neutraceuticals known as palladium lipoic complexes. He insisted we look at a video which showed the testimonies of twelve people whose lives were saved by this wonder medicine. This guy was acting exactly like some high pressure used car salesman and I was rapidly developing an intense personal dislike for him.

Next, he really ticked me off when he asked me what I was thinking because, "I can also read minds, you know." I felt like saying, "Well, right now I'm thinking I'd like nothing better than to jump over this desk and smash my rather large fist right into your earnest, smiling face."

I managed to control myself and said, "Well, if you can read my mind, why don't you tell me what I'm thinking?" He said he figured I was feeling skeptical about the whole thing and I assured him he was bang on there. What I really wanted to know was what Margo's odds were if she went on this treatment and what the total costs were going to be.

After some more beating around the bush, he admitted that the cost of the recommended treatment would be $5000 for the first two months, then $2500 for a smaller dose for the next two months and after that it would be $625 per month for the next year. On top of that would be the other supplements at a cost of just under $400 per month. This was a far cry from what we'd expected and it really made us sit up and think.

He assured us he was only going to charge us his costs to bring this expensive medicine in, because "I will make my money off the patients you'll send me when they see Margo walking around cancer-free six months from now." I felt like calling him a rotten liar. Who does he think we are that we would believe for one minute that he isn't going to take a huge markup? He knows we are desperate people grasping at straws. He knows we have no choice but to go ahead with the program no matter how skeptical we may be.

By the time we got back home we'd decided to give it a shot and hang the expense. We think nothing of spending large amounts of money on machinery and it would be very stupid and cruel to turn our backs on something that could possibly give Margo years of good health."

On Friday, August 20, we went to get Margo started on her new treatment. We listened for an hour and a half while the doctor lectured us on what we could and couldn't eat from now on. We wrote a cheque for $5400 for the first two months and he promised he'd have the medicine there on Monday.

That afternoon, Donna and her two kids arrived from Edmonton. Margo brightened right up when she saw her grandkids. A short while later Tracy and Dylan arrived for the weekend and we all got to see the little guy take his first four steps that evening.

On Sunday, all seven of us piled into Tracy's van and went to church. Bill spoke a very good sermon admonishing us not to pre-judge people who we just met. Perhaps I'd been judging this new doctor before he had a chance to prove himself. I promised myself to lighten up and wait to see how the treatment turned out.

On August 25, Donna and the kids left for home and our house felt silent and empty without them. It had been great having them here. Donna worked like a trooper cleaning house and cooking for us and Daniel and Becky had been very helpful as well.

The last day of August found us back in the city for another check-up from Margo's oncologist. The tumour markers continued to rise and he was totally baffled as to what to try next. The only thing that seemed to knock the cancer cells down was the chemotherapy that had been almost killing her. It had been four months since her last treatment. He did say that her liver was functioning as well as a perfectly healthy one and we were grateful for that.

That evening we drove to the airport to pick up Kevin who was coming to visit for about ten days. It was wonderful to see him looking lean and healthy. We hadn't seen him for close to two years so we had lots of catching up to do.

It had been a long and bitter summer and now our only hope was this alternative therapy. If all went well we could be seeing a big change for the better by Christmas.

NINE
Our Last Autumn and Early Winter

We began September, 2004 with new hope. While I still had grave doubts in my own mind regarding this new treatment, Margo was ecstatic and her happiness rubbed off on me. It did seem as though she was on the mend and feeling so much better. I knew in my heart that part of her well-being was a result of her body ridding itself of the last of the chemicals from her chemo treatments. Her sore feet and fingers were healing and she was able to walk without a lot of pain for the first time since her treatments began.

It was wonderful having Kevin home again. He left us at the age of seventeen to find work on the west coast. At times it seemed as though he was just a very good friend who we didn't see very often. He'd brought a cooler of frozen seafood with him. On his first weekend home, we invited Steven and his family out to the farm to enjoy the wonderful feast Kevin cooked for us. We sat around the deck and talked about getting our whole family together for a photo. It had been years since we had all the kids together in one room with us but there didn't seem to be any way we could make it happen.

Margo loved having her big boy here too. They went out together several times to play Bingo in town—a tradition they shared during his visits home.

"Margo and Kevin came home very happy because Kevin had won a pot for the second week in a row. Most importantly they had fun together which was the whole idea. None of us knows where we'll be or what we'll be doing a year from now but I guess that's the way it's meant to be. We have to keep on hoping and praying for the best. It's so wonderful that all you guys made the effort to come home and see her. It really meant the world to both of us."

On September 16, I drove Margo to the Eriksdale hospital where she was given the intravenous treatment to build up her bones. It was a long three hour session in that tiny emergency room, a cheerless place with no windows. To help pass the time, I brought my guitar and sang several songs for Toots and her daughter, Janet, who were in the room around the corner. Toots told me she was raised on a dirt farm near the town of Narcisse, Manitoba. At sixteen, she was milking twenty-five cows by herself while her family was in the hayfield. I sang one of my favourite songs, The Old Dairy Cow, and she surely enjoyed it.

We'd been enjoying having Dot visit for a couple of weeks. On September 25, she left for her home in Bellingham, Washington. I'd been writing my diary letter to her at the time.

"I don't know if you realized it or not but when we were hugging and kissing each other goodbye, I almost broke down and cried. I'm becoming quite a cry-baby in my old age, after all these years of pretending that strong men don't cry. I love you lots, like I love all of my family. The older I get, the more I realize how precious you all are to me. It's so nice to stay in touch by letter or phone calls.

When one is young and strong, some things don't seem to be so important but as we get older and our health starts to fail we begin to realize the true value of family and good friends. People who are foolish enough to think they are strong enough to go it on their own will most certainly some day find themselves dying a sad and lonely death."

On September 28, we drove to the city for Margo's blood tests at the Cancer Clinic. We brought a carload of fresh vegetables to give to various friends along the route. The next day we drove back down that long highway to see Margo's doctor and get the test results.

"Dr. Brandes was all smiles when he came in and he said the news was fairly good. The blood tests were giving mixed signals with one tumour marker going up a bit, which isn't good, while the other marker had gone down quite a bit. The liver functions which he was quite pleased with last time were now better than ever. He isn't sure what to make of it but he is hoping the chemo pills he prescribed last month are working.

Margo and I smiled at each other but kept our lips zipped about the alternative therapy. We don't plan on telling him about it unless she goes into complete remission. He let us know from the start that he wasn't in favour of such things and we had respected his wishes until he told us there was nothing more he could do for her.

This is the first good news we've had for a long time. Actually we weren't all that surprised because Margo has been feeling so much stronger over the past few weeks."

On October 4, we attended church in the morning, and had lunch in town before coming home for a good nap. Here is an incident that happened that afternoon.

"I finally dragged myself off the couch and went hauling hay. I finished hauling the last two loads along the lakefront west of here and then started in on the low-lying Crown land to the southwest. I started picking bales along the lakeshore and gradually worked my way back toward higher ground. I got greedy by loading too many huge bales and got the tractor stuck in a low spot. I didn't spin the wheels down, just shut the engine off and phoned Margo to see if she felt up to coming out with the loader tractor to give me a pull. She was there within fifteen minutes and soon had me pulled out without even having to unload the bales."

For years Margo had told me she sincerely wished she would never have to tow me out of a wet spot again but neither of us realized this would be the last time she would ever do that.

October 9, 2004 brought the annual All Canada Goose Shoot put on by the Lundar Elks Association. This hunt has sixteen teams of four hunters, each competing to see who can bag the most geese. If more than one team shoots their limit, the birds are weighed to see whose geese are the heaviest. Ron Heroux and I were acting as guides and this year I had two fields entered in the hunt. One team of young fellows from Lundar drew my barley field. Ron and I guided another Lundar team who were hunting in our pasture land south of the barnyard.

The goose gods were smiling down on us. The team we were guiding shot their limits before eight o'clock. They won first prize while the team in the barley field took second place.

Margo and I attended the dance in Lundar so I would be on hand to accept the brand new pump shotgun that is awarded to the land owner of the winning team. Margo's sporting blood came to the front when she bought a handful of tickets for the large silent auction draw. Her name was the first one picked and she walked up front with her cane to choose a beautiful case for my new shotgun.

Jayden Robert Watson arrived on the evening of Friday, October 15, a brand new son for Steven and Linda and the eighth grandchild for Margo and me. The birth went well and Steven brought Linda and the new baby home on Sunday.

We attended a dance in Eriksdale that weekend. Margo was able to go without a hat or wig for the first time since I'd shaved her head about a year and a half earlier. She looked quite cute. Her hair hadn't come in curly as we'd been hoping but it was real hair and she was proud of every wisp. Her energy level was climbing all the time and we danced until I was tired and wanting to go home. She made me stay until the dance ended to make me suffer for all the early morning hunting I'd been doing. That was alright by me. She was having such a good time, she didn't want to go home and who could blame her?

Monday morning brought the beginning of the muzzle loader season for deer. This meant many hours sitting in a blind watching a game trail while my wife was left home alone to amuse herself. She'd been feeling so much better we were starting to believe this alternative treatment really was going to kill every last cancer cell in her body. I did my best to convince her to hunt with me. She told me not to push my luck as she wasn't feeling quite that spry yet.

On October 22, we harvested a large crop of carrots in our garden. There was rain in the forecast and we wanted to get the job done before winter set in. Margo was feeling quite strong as she worked alongside me.

That was another last job done together, although we didn't realize it at the time.

A few days later she baked seven pies and it did seem like life was returning to normal. My waistline was starting to expand now that I was eating Margo's good cooking again but oh, those pies were good!

We continued going to dances on weekends. We loved being with our friends and listening to the music even though we didn't dance that much. Our friends continued to rally around us and give us much needed support. Margo said she never knew how many friends she had until she got sick. That was exactly how I'd been feeling.

On October 25, we were back in the city to see Dr. Brandes.

"He said her tumour markers are coming up again after a drop in one of them last month. He was really puzzled because her liver functions are doing better than ever, so again we have mixed messages. He suggested leaving her on the chemo pills for another month but Margo and I both got the impression he doesn't hold out too much hope for her.

We battled rush hour traffic to get out of the city but were out at John and Pat Fjeldsted's house north of Selkirk before supper. Pat made a beautiful supper and we ate until we were stuffed. We talked until nine and then had to head for home."

That visit with John and Pat was another last time come and gone for us without us even realizing it. How fortunate it is we can't see into the future or it would take away the delight of the present moment.

On October 30, we drove to Lundar to pay last respects to Toots Paulson who had lost her battle. She'd been one very brave lady and my only wish was that I might've met her earlier in our lives in happier times.

On November 4, Margo asked me to wash all the windows on the house before I went out cutting wood. That actually worked well in my plans because I knew a bunch of her close friends were going to surprise her with a party the next day. I'd been wondering how I was going to spruce the place up without being too obvious.

"Nov 6, 8:45 AM

Today is Margo's sixty-fifth birthday. A year ago we weren't sure if she would live long enough to see it but she is still here on the right side of the clouds and doing quite well.

Before going out to work on the wood pile yesterday morning, I vacuumed, washed the kitchen floor and cleaned up some of the clutter without arousing Margo's suspicions.

We had lunch at noon and there was time for my nap before our friends started arriving. Margo slept for awhile too but she woke up in time to see the first car pulling into our driveway. The ladies had met half a mile up the road to make sure they all arrived at the same time. It was really fun for me to go along with the gag by going to the window to see a whole parade of cars and trucks pulling in. Margo was really guessing as to what was going on and at one point she exclaimed, "Holy crap! It looks like a funeral procession." Then, of course, we dashed to the kitchen to greet the guests.

The poor gal was overwhelmed when she started recognizing her close friends. When the hugs started, the tears were welling up very close to the surface.

I put all the extensions in the kitchen table to make room for all the food and the dozen yellow roses they'd brought. It wasn't long before everyone was standing singing Happy Birthday to the special girl. I'd just read that people with a strong support group do much better than others when it comes to fighting cancer. Margo has so many good friends who love and support her. Perhaps that's why she's doing so well."

This was the last birthday party she would have and it was so special. We celebrated Margo's actual birth date by cutting and wrapping my deer which had been hanging in the shop. Margo was pitching right in like she had in the good old days when we were raising our family on venison. She even washed the meat saw for me. That evening we attended a dance at Meadow Lea Hall. It was well after midnight before we headed home.

Winter arrived and the cold weather triggered a desire to go west to visit Kevin and his family. When Ruth volunteered to look after the farm with the help of neighbours, plans were made to fly to the coast in January.

On November 23, we were out of bed at five in the morning for another trip to the city to see Margo's oncologist. The highways in the Interlake were sheets of ice but we managed to be at the clinic when the lab opened at eight o'clock.

Our visit with Dr. Brandes went about as we'd expected. He said the chemo pills didn't seem to be helping because her tumour markers were continuing to go up. He told us she may as well quit taking them because at times they can cause the cancer to spread. This was exactly what the alternative doctor had told us. Margo's hip had been causing her lots of pain and she was using her cane at all times now. The doctor ordered a series of x-rays to be taken before we left the hospital so he could be sure the cancer wasn't eating a hole in the bone. He didn't want her hip to collapse and neither did we.

November 28 brought a major change into our lives.

"Tracy and Dylan stayed home while Margo and I went to church. We hated to run off on them but I'd been asked by David Ives to bring my guitar and lead the singing, so I couldn't back out. It was a good thing we went because we were there in person to hear the bombshell Bill laid on us. He's been hired as a full time pastor in a larger church in Ashern.

We are happy for Bill but sorry for ourselves. He's a wonderful preacher and becoming so much in demand for funerals and weddings that he's being run ragged. He told the church people in Ashern he will continue singing at the hospitals so I guess I won't be losing that part of my life, at least not yet. We are going to miss Bill."

On December 3, Donna and kids arrived safely in time to sample freshly baked bread and cinnamon buns. Margo had been feeling under the weather with a cold but she shrugged it off long enough to bake this treat for our special visitors.

The next morning Donna was off to the city with Margo and Becky on some sort of mysterious errand. The weather wasn't good and I tried to talk them out of it. When an ice storm arrived that afternoon I phoned Margo and caught them shopping at Sears. I told her I was concerned about the state of the roads and advised her to head for home. Her answer was that they would get here when they got here and to quit worrying.

The ladies arrived home safely later that evening. The object of the trip was so Donna could surprise her mum with a gift of a new queen-sized bed. I hated to see the kids spend their hard earned money on something like that but had to admit it was going to be nice to have a wider bed.

All fall I'd been having pains in my stomach coming from an area near where the blood clot had been in my liver. I hadn't said anything to Margo and didn't intend on saying anything unless the pain got extremely bad. As time went on, it was starting to wear me down.

"Yesterday was really lousy for me mainly because I wasn't feeling well, mentally or physically. Ever since the blood clot episode last winter I've never really felt 100%. Lately I've been cutting wood. Every now and then I lift a heavy block of wood and feel the pain again. This past week it's worse than ever. Put it all together with Margo's health problems and no wonder I get feeling down. The last thing I need right now is to fall into a depression, so I'll have to speak to myself sternly and smarten up.

Last night we took in the annual Christmas supper and dance at the Lundar Legion where we are members. The roast beef supper was wonderful and the dance that followed had a good band playing. We didn't stay until the end because Margo was getting tired. We are both thankful we can still get out together."

Donna and the kids went home on December 7. The next day we attended a funeral for Marlene Douglas. She'd been diagnosed with lung cancer in May and she was gone already. Seeing how quickly she had succumbed to the disease really put things into focus for us. We realized how lucky we were to have Margo still here and actually doing quite well. While sitting beside Margo during the service, I marvelled again at the courage she was showing by attending a service like this when she knew the next one could be her own.

Once a month we drove to Eriksdale hospital where Margo was given the intravenous treatment to keep her bones strong. One young man there had really captured Margo's heart.

"December 15

Our good nurse Werner was waiting for us, trying to convince us we were late. It didn't take him long to get Margo comfortable on a cot with fluid running into her arm. Her left leg has been really sore below the hip and we are imagining all sorts of bad things. If there is another cancer spot we don't want the bone to collapse so these treatments must continue. It's more than a two hour procedure so I'm just there for moral support.

Werner visited with us for the last half hour so the time went by fast. He's a farmer as well as a full time nurse but he'd much rather be farming full-time if he could afford to do so. His wife is totally sick of the struggle and is urging him to sell the cattle and get clear of the heavy workload they are under. It's a sad story that is repeated over and over on the farm scene these days."

On December 17 we were off to town again for Margo's check-up with Dr. Steyn. It was so wonderful to have a family doctor who really cared about us.

"Margo had her check-up and Elzette called me in for a conference. She is worried about Margo suffering pain in her hip. She wants us to ask Dr. Brandes about having radiation on that hot spot for pain relief. I assured her Dr. Brandes is already prepared to have radiation done when it becomes necessary. Right now Margo doesn't have much pain unless she is moving around and she sleeps like a baby at night."

On December 20, 2004 we got to the Cancer Clinic by eight o'clock.

"We were dreading our visit with Dr. Brandes but at least we know where we stand. There is nothing he can do for Margo now. He's willing to do whatever we want, even if it means putting her back on chemotherapy but she told him she'd never do that. I asked him straight out how long we might expect her to live. He has no way of knowing for sure because there are so many variables but it could be weeks or maybe months. I was hoping he would say years but he didn't. He said it's definitely time to get our affairs in order and at least we'll have a chance to do that.

He suggested radiation on the bad spots on her bones but we told him that will be a last resort. He wanted to know why and, of course, we couldn't tell him that our alternative doctor is dead set against radiation because it weakens the immune system so quickly. So far we haven't told Dr. Brandes anything about the other treatment and feel no need to do so now. He told us there is no reason to come back to the Cancer Clinic unless we feel she needs radiation for the pain. He gave Margo a long, tight hug and we could see tears in his eyes.

Hearing this horrible news really knocked us for a loop. The harsh reality set in and it was a long, sad ride home.

We got home after dark and were very thankful that Steven had done all the chores. We got several phone calls for Margo but I did all the talking because she wasn't feeling up to making conversation.

This is going to be a sad Christmas for us but we'll love up our kids and make the best of things. It most likely will be the last one we share together here on earth.

I have such a range of emotions running through my head; it's impossible to try and explain them. I think the main emotion I'm experiencing is fear of being left alone after Margo passes on. She's been my rock for so many years, sticking with me through thick and thin. There isn't a spot on this whole farm where she hasn't worked side by side with me. There will be lots of happy memories everywhere I look but that won't be the same as having her here with me.

Another strong emotion churning deep inside me is that I must face the fact there is not a thing I can do to make this nightmare go away for her. All these years I've been her protector but how can I protect her from this evil disease? At least we know this is coming and we have a chance to make plans and set our house in order.

It's strange what we can face when we have no choice in the matter. It's so very weird to sit together at the breakfast table talking about things that have to be done while she's still well enough to do them. Margo is so brave but when I comment on that she says, "What choice do I have?"

My biggest wish is that she'll come to accept and love God before her time comes. It would make this a lot easier to face if we knew we were going to see each other again after we die. I'd love to spend eternity together with her in Heaven."

Christmas came and went. Margo's health continued to remain about the same, but she was starting to complain about severe pain in her lower back and hip area.

New Year's Eve was spent quietly at home with friends, Anne and Arnold Larson, playing our favourite Sequence board game. With all the laughter and good vibes flowing, it was hard to imagine anything was wrong in our lives.

We were looking forward to our trip to the west coast to see Kevin and his family. In the meantime it was business as usual to make sure nothing was missed as far as Margo's health was concerned.

"January 12, 2005

Yesterday Mum got her intravenous bone builder treatment. I waited with her while Werner got her hooked up and comfortable with a warm blanket. A few minutes later Werner put a man on another cot behind a curtain in the same room. We recognized Arnold Kirby's voice so I spoke to him saying, "We've got to stop meeting like this, Arnold." He recognized my voice and we had a good chat which helped pass the time.

Arnold is full of cancer which blocks off his stomach so he throws up everything he eats. He will never be free of this cancer and he's had at least two operations to open the blockage in his stomach. Right now he has to come in once a day to have the dressings changed. He also has terrible pain in one hip like Mum has and they give him some kind of pain killing patch. He is much like Mum in the way they are both so positive in their attitude. It was funny to hear them laughing and joking about trading hips. Of course they both wanted to end up with two good hips. Arnold isn't giving up his battle for life and plans to go in for another round of chemotherapy soon."

On January 22, we flew to the west coast. Margo's hip was really bothering her so we asked for a wheelchair so I could wheel her down the corridor to the plane.

We had five hours to spare before catching the evening flight to Powell River and were overjoyed to see two of Bertha's children, Colleen and Doug, waiting for us in the Vancouver airport. We had something to eat and it wasn't long before we were joined by Bertha's youngest son, Dwayne, his wife, Lorraine, and their daughter, ShayLyn. The time passed too quickly and before we knew it, we had to board the plane.

Kevin, Natalie, and the kids met our plane and Steven and Kristen were each carrying a bouquet of flowers for their Grandma. That was almost too much for her and she was fighting back tears as she hugged all her dear ones for a very long time. Leaving the airport we drove through the dark streets of this small town where we'd spent the first three years of our marriage. Kevin and Nat had bought a house in Wildwood with a large yard. This was the first time we'd been there and it was great to see how the whole family was so proud of their home.

That week was wonderful for all of us. We became reacquainted with the grandkids who had grown so much since the last time we saw them. Kevin and I were on the ocean every day lifting crab and prawn traps. We ate seafood for five days straight. Kevin put me to work helping him put a ceiling in his shop as well as other renovating projects to make things more convenient for him. I wasn't feeling very well because that nagging pain was back in my stomach. I had no energy and after working for a few hours, I would escape to the house feeling completely wiped out.

While we were staying there we made a decision to sign up for the Internet as soon as we got home. We'd been resisting this move for years but when we saw how it brought the whole world to your fingertips, we decided to make the plunge into the modern world.

A highlight of the trip was an evening with our nephew, Bob, and his wife, Francine. We ate a late supper in a restaurant on the edge of Powell Lake and talked for several hours.

We made sure we visited all our old friends. That was hard to do because we all knew this would be the last time they saw Margo. It was over thirty years since we'd met these people but we'd never lost touch.

On the evening we were supposed to fly out of Powell River, the airport was fogged in solid. We could've taken a chance on the morning flight going out but didn't want to miss our connecting flight back to Winnipeg. Next morning we all got out of bed at 4:30 AM and Kevin and Nat drove us to the second ferry at Langdale. We walked on that ferry and I got a wheelchair for Margo so I could wheel her to the parking lot where we were met by Margie and Larry.

They drove us to the airport and we visited until it was time to catch our flight home. That was a bittersweet visit because, they too, knew this was the last time they would see Margo. We tried to make happy conversation but the tears glistening in my little sister's eyes almost brought on a real cloud burst from me.

Back home life settled into a comfortable routine as I tried to do most of the housework with Margo pitching in whenever she felt able. I was struggling with the computer and was starting to question the wisdom of signing up for the Internet which was bound to be even more complicated.

"February 5, 2005

I spent several hours fighting with this stupid computer. Later I managed to do something I could handle when I stripped our bed and washed the sheets and pillow cases. Margo must've been inspired by that because later she washed two loads of laundry and did a bunch of ironing as well."

On February 9, I drove Margo to see Dr. Steyn.

"Margo has been getting lots of pain in her lower back and abdomen, along with lots of pain in her left hip when she tries to walk. Elzette is going to line her up with an emergency CT scan to check it out. In the meantime, she prescribed patches to control the pain in her leg. The patches are $58 for five which will last for fifteen days.

When our doctor came out of the examining room with Margo she came straight over and gave me a big hug. I don't know whether she thought I needed comforting or if she was the one who needed the hug. She has a very compassionate heart and worries so much about her patients."

That night Margo came along with me to join Bill and Joanne for singing at the hospital. I was so glad Margo was beside me to hear the rather remarkable story that Bill related to us.

"Bill had conducted a service at Faulkner that afternoon and while he was there, a young lady introduced herself. He said she looked like a teenager but she told him she was the mother of three kids already. One evening last summer she was lying on a gurney in the Eriksdale Hospital emergency room after having suffered a miscarriage. She was having uncontrollable bleeding and was being prepped for an ambulance ride to the city. She was lying there wondering if she was going to live to see her kids again when she heard Bill and I singing in a nearby room. The room is right next to the emergency area and this young lady heard us sing three hymns before we moved on down the hall. All the way to the city in the ambulance she clung to the words of those songs and she thinks it helped her survive her ordeal. She told Bill she wanted to thank him and he told her he'd pass her thanks on to me.

It makes me feel good to hear a story like this and to know our efforts are worthwhile."

Two days later we went to Selkirk Hospital where Margo was given a CT scan on her abdomen. This scan was preceded by her drinking a litre of water as well as two large cups of oral contrast water. By the time the scan was done she was feeling quite ill from drinking all that fluid but it was such a beautiful day she cheered right up on the way home.

That evening brought good medicine when Tracy and Dylan arrived to spend the weekend with us. Grandma could still give the kind of hugs that little kids love. Tracy spent some time teaching me how to email even though I wasn't hooked up to the Internet yet. My life was about to change in a hurry with the help of modern technology.

TEN
Learning the Email Game

The calving season arrived with a couple of calves born on February 15 during a spell of very cold weather.

Two days later, Dallas Moman arrived to set up my computer and to get us hooked up to the Internet. It wasn't long before Margo and I were tapping out tentative messages to a group of friends and family on the coast. I was very nervous using the Internet at first and Margo only tried it to please me.

Here's an email that I sent to Margie.

"Dear Margie,

It's been only three days since we got hooked up to the net but already I'm feeling a lot more comfortable with the whole deal. Today I was able to read my three-in-one manual and make enough sense of it to be able to scan a couple of pictures and send them away. Sometimes I'm amazed that my old grey matter still works well enough to take on something like this.

We went to a dance in Lundar last night. Margo isn't able to dance and that really irks the heck out of her. I had a few dances but don't like to leave my sweetie sitting there by herself. Actually she got plenty of company with lots of friends coming to visit at our table. On Sunday she got three phone calls from Arizona and today she got one from Texas so you can see she has a real good support group.

She's running low on pills so I think we'll try to go to the city tomorrow to stock up. I'll ask Steven to check cows for us. There are four new calves already."

The next day we sneaked away from the cows long enough to make a flying trip to the city to see Margo's alternative therapy doctor. Margo could barely hobble into his office. There she sat with dark bags under her eyes and her face a pasty, grey colour while he raved about how wonderful she looked. In January we had written him a cheque for $4,300 and now it was time to pay the piper again. It was so plain to see she wasn't improving, no matter how hard he tried to convince us that she was. To make matters worse he'd raised the price of her medicine from $1,000 per bottle to $1,250.

"The doctor seemed happy with Margo's general condition although he wasn't pleased to hear she is now on the pain patch. We'd brought the container in with us and he read the list of ingredients and found they contain opium. The patch has made Margo really drowsy and she's been sleeping lots during the daytime and all night too. Another bad side effect is suppression of appetite. I know she hasn't been eating well since she started using the patch.

He suggested she discontinue the patch and ask for a shot of cortisone in each hip. I suppose cortisone has side effects too but he must figure it's the lesser of two evils. I do know Margo has been doing well on this alternative therapy treatment. If she didn't have constant pain in her hips and lower back she would be just like her old self."

What with learning the computer game and calving cows I was feeling my age at times but tried to keep my sense of humour as this next diary entry shows.

"This getting old isn't all it's cracked up to be. I seem to be getting deaf as a post and that causes all sorts of problems. Margo tells me I have selective hearing but I really am getting deaf, no fooling. Here's an example of an ordinary conversation around here.

Margo... "I was talking to Mabel this morning." Dan... "Eh?" Margo... "I said I was talking to Mabel this morning." Dan... "What?" Margo... "Oh, never mind. You're not listening anyway." Dan... "No, really, I was listening. I do want to know what you said." Margo, looking very annoyed, says, "I just told you that I was talking to Mabel this morning." Dan... "Oh. What's the matter with the table?" Margo takes a deep breath and yells, "I told you that I was talking to Mabel this morning." Dan, feeling slightly hurt, replies, "Oh, really. Well, you didn't have to shout at me." And so goes a day in the life of the Watson family."

Actually my hearing wasn't nearly that bad and I was just trying to put something on paper to make Margo laugh. More and more I found myself being silly and playing the fool to try to make her feel better and help her through the day.

One evening on the phone I was talking to someone and telling them how doubtful I was about that doctor and his alternative therapy program. I was forgetting that Margo was sitting right there taking in every word. When I finished my call, she looked at me and said, "Do you have any idea how that makes me feel when you talk like that?" Before I could answer, she spoke again, "It makes me feel like you have absolutely no hope for me."

That gave me a lot to think about over the next few days and I was very careful how I chose my words from that day on.

By the end of February, Margo's pain was becoming unbearable and the pain patches didn't seem to help at all. We reached the point where we were seriously considering medicinal marijuana. I'd done some research on the Internet and we'd discussed it with Dr. Steyn who felt it might work as well as anything else on the market. We knew she could get pain relief by using morphine prescribed by her doctor but we also knew once she started on that, it would be the beginning of the end for her. Morphine would dull her appetite and cause her to lose interest in her surroundings and she'd end up needing more and more. I hated to see her suffer but at the same time I didn't want to lose her any sooner than necessary.

I also researched the Internet to find some place to buy the expensive medicine we'd been buying from the doctor in the city. Imagine my surprise when I found out the real cost.

"March 2, 2005

We looked up some information on the medicine Margo has been taking. We've been paying $1,250 per eight ounce bottle which is enough for twelve days. We found exactly the same product in the same bottle advertised for $200 per bottle. Even with exchange and duty added, there would be a tremendous saving if we were to buy direct from the supplier. I can't wait to have a chat with our dear doctor."

That evening Margo came along with me when I sang at the hospital. Dr. Steyn was working her 24 hour shift on ER and she took us to the TV room where we worked out a plan of action. She was going to try to get a capsule form of cannabis for Margo but it was so new, most druggists weren't aware of it yet. These pills would supposedly increase her appetite and lighten her mood. She'd be aware of her surroundings and alert, not like the effects of morphine.

The next day Elzette told us she'd secured a prescription for the cannabis pills. The cost of $900 per month would be covered either by Manitoba Medicare or the Cancer Clinic. That took a load off our minds because we were wondering how we were going to swing it.

We had the new pills that afternoon so Margo was able to get started at once. She was to start off with one pill a day for three days and then two pills a day for seven days and so on until she was on a full dose of four pills a day. Margo took one pill at four o'clock and by eight-thirty she was feeling nausea and dizziness which are some of the known side effects. She went to bed and had no more trouble.

As I became more familiar with emailing, the list of contacts in my address file was growing. I liked staying in touch with friends and family who were all concerned about Margo's health but it was becoming increasingly difficult to send individual emails every day. That problem was partially solved when I formed a small group named Letters. This meant I only had to write one letter to the whole group. I was totally enjoying the experience of sitting at my desk and opening my heart to loved ones in a way I'd never been able to do before.

I kept searching the Internet until I found the company in the States where our city doctor was getting his supply of expensive medicine. On March 8, I phoned and had a long talk with a very helpful lady. We worked out the price and found that after all expenses, the price per bottle came to slightly more than $200 in Canadian dollars.

The lady asked me where I'd been getting the medicine from and when I named the doctor, she told me they get wonderful phone calls all the time from people who had been dealing with him. That information made me wonder whether those people were, like me, trying to get a decent price instead of paying his rip-off price. I told her I'd contact her within a few days to place the order.

When I hung up the phone I was in such a rage I was ready to drive straight to the city to have it out with that con man but Margo wisely talked me out of such foolishness.

"Later I phoned Customs and talked to a girl who assured me it would be no problem for us to get the medicine by mail from the States. There is no duty on it as it is listed as a food supplement and the package would come straight through to our post office here in town. Can you guess where we'll be doing our buying from now on?

I've been suspicious of this doctor from day one, especially when he kept going into his spiel about how drug companies are taking advantage of helpless people. Talk about taking advantage! He's been doing a good job of wiping out our savings. It's not that I don't think he is a good doctor. I do believe his program is working for some people. It's just that he is ripping us off and trying to keep us on the hook for as long as possible."

I'd read stories about people dying of cancer who had fallen for the same kind of deal we'd found ourselves caught up in. I'd wondered how anyone could be so utterly stupid that they would wipe out their life's savings while traveling to foreign countries in search of the cure they couldn't find in conventional medicine. Now I knew the answer to that. Desperate people take drastic measures.

"March 11, 2005

Margo's pain is getting worse every day and it sure is wearing her down. The cannabis pills are making her sleepy all the time and she's still only on half a dose. I feel so helpless. I can't do anything for her except the cooking and housekeeping and trying to make her comfortable. I wish there was something that would work for the pain."

On March 14, we told Dr. Steyn we weren't happy with those new cannabis pills. Margo had been completely groggy and disoriented to the point where I was afraid to leave her alone for fear she would try to use the stove and possibly set fire to the house. The pills hadn't eased her pain and in fact, it was getting worse. Elzette thought it was time to try another approach. That evening I sent an email to Tracy describing the latest development.

"Hi Tracy,

It's time to have the hot spots in Mum's spine treated with radiation. She will take a five day treatment and then no more until the pain gets bad again. The radiation will shrink the lesions and lessen the pain. In the meantime she's been switched to morphine pills to take as needed. She took the first one about four o'clock and within twenty minutes the pain was gone. She's visiting with Bertha right now without pain showing in her eyes for the first time in a month. The morphine is highly addictive and will cause her intestines to slow down so she's going to try and get by with as little as possible for as long as she can."

Bertha had just arrived for another visit. The timing was right because the morphine pill let Margo relax pain free and the sisters chattered away all evening.

On March 16 at our Wednesday night jam session in town, we surprised Ruth with a huge cake for her sixty-fifth birthday. Ruth had been such a help to us through good and bad and was always there when we needed her. She got a bit teary-eyed when she saw the cake but managed to pull herself together and make a little speech.

Here is part of an email I sent the next day.

"There's not much to be happy about these days but I'm still totally thankful for every day we can be together. It really is wonderful having Bertha here. I wish she was staying all summer because we love her so much. It's a real worry to me to have Margo alone while I'm out working but she doesn't seem to mind. She's doing better since getting off the cannabis pills and she gets by with only three morphine pills (2mg) in twenty-four hours.

It seems strange to talk about issues like this but because it all came on so gradually, it has become just part of our every day conversation. I hope she can get in for radiation as quickly as possible so the morphine dosage doesn't have to be increased.

I really love this Internet thing and wonder why I held off so long. I was afraid to try my hand at it because I didn't think I was capable. This has done a lot to restore my confidence in my abilities. Love Dan"

Our children were taking the news hard and Tracy confided in me that she'd broken down at her workplace several times in the past couple weeks. Here's an email I sent to try to comfort her.

"Hi Tracy,

Don't be ashamed about losing it in front of your boss. Sometimes we have to let go and let the tears fall where they may. I've been losing it once in awhile over these past two horrible weeks but am happy to say that Mum is a lot more like her old self today. We have an appointment with a radiologist at the Cancer Care Clinic on Monday.

Don't be too sad, my girl. You are a lot older now than I was when I lost my dear mum and I never even had a chance to say goodbye. I know that by the time your mum passes on, I will be thankful she is at peace at last. I hope God enters her heart and she becomes a believer before she passes on. It breaks my heart to think we might never meet again in the next life.

I guess it will all be clear to me in the end. In the meantime I'm going to try to live a clean life, hoping I move on with a clear conscience to whatever awaits me. Love, Dad"

Here's an email written March 21 labeled, "Margo Update".

"Hi Guys,

Margo saw a lady doctor at the Cancer Clinic today who will administer radiation treatments to shrink the cancer spots on her hip bones and spine. We came out with an appointment for a CT simulation where they do mapping and marking to prepare for the radiation. This scan takes place on Thursday afternoon but we're not sure when the radiation will start. There will be one treatment a day for five consecutive days.

We dropped Bertha off in Selkirk where we met up with their sister, Florence. Tomorrow morning Flo will drop Bertha at the airport. We're sure going to miss having Bertha here but we know she's anxious to return to her own family. It has been wonderful medicine for Margo having her big sister here for a week. Love, Dan"

On March 22, six bottles of medicine arrived from the States at a total price of around $200 per bottle.

That afternoon I butchered a big steer, loaded the sides of beef in the truck and delivered them to a small butcher shop at Moosehorn, to be aged before being cut and wrapped. Normally I'd have done the whole job myself but felt it was too much for me this year. I finished off my journal letter that evening.

"It was getting late when we got home and I still had to do all my evening feeding. I was so tired I could hardly drag myself around the yard but kept at it until the last chore was done. Then I still had to cook supper for the two of us. I realize now how much work Margo did all those years, raising kids, milking cows, slopping hogs, cooking for the gang and cleaning house, etc. They say a man works to dusk from dawn but a woman's work is never done and I believe that.

Margo doesn't have much appetite but if I make a good meal, she'll force herself to eat some of it. I hate to see her going through so much pain. Hopefully she'll soon have the radiation and we pray it will help. Love Dan"

The next morning we met with the radiologist. It was a long day and that night I sent an email to my Letters group.

"Margo Update March 23, 2005

Hi everyone,

They took Margo in and scanned her, marking black lines in the shape of a cross from her rib cage to her hips. Then we waited close to an hour before they gave her the first radiation treatment. Dr. Vijay knows we are from out of town and set us up with the first treatment today. The technician told Margo not to expect any improvement in her pain from this first treatment. She goes back on Saturday and again on Tuesday, Wednesday and Thursday. She was in lots of pain tonight from traveling and is sleeping already. I'll keep you all posted. Thanks for all the emails that were waiting for me when I got home. I'm too tired to reply tonight. Dan"

Word was spreading among our large family that we were on the Internet and our email address list was growing. It was such a comfort to get even a few lines letting us know people were thinking about us and cared enough to write. I was still trying to keep up with my journal letters because some of my family members weren't on the Internet but it was becoming increasingly difficult. Here's an entry from my letter to Donna where I was trying to catch up.

"March 27, 9:20 PM

There hasn't been time to write on this letter for three days. Friday morning Ruth baked us a batch of cheese bread and some delicious cinnamon buns. Tracy came out by herself because she didn't want us to catch Dylan's cold. She was a wonderful help all weekend as she pampered Mum, even giving her a foot massage and pedicure. She also cooked meals and cleaned house, in between teaching me new stuff on the computer.

That night the cows kept me hopping until almost four in the morning. I had to be up by six o'clock to drive Mum to the city for her second radiation treatment. They took her in right away and the two technicians invited me in while they got her ready. She lays flat on a big table while they line everything up using laser lights. The three of us went into another room and watched her on two television screens. We could see Mum lying quietly while the machine revolved right around her. The actual treatment took just over two minutes.

We headed out of the city but hadn't got far when Mum got violently ill and started throwing up all over herself. It was so unexpected that we never had anything for her to chuck up into except a leaky plastic shopping bag. I pulled over away from the heavy traffic and emptied out a bag of buns. Her nausea soon passed but it was a long, sad ride home.

The car was shimmying from ice or mud inside the rims. I got thinking I should stop in Lundar to get the wheels rotated and the rims washed. Mum phoned Ellen Steinthorson to ask if she could stay with her while I was at the tire shop and Ellen said she'd be waiting for her. When Mum heard Ellen's kind voice she broke down and started to cry.

Next she phoned Tracy and then she really lost it. She was so upset I decided to just bring her straight home. It seemed everything came to a head on that miserable ride home and that caused the breakdown. That is only the second time she's really cried since this nightmare began. She's the strong one in this pair. At home she settled down okay with Tracy caring for her and things became as normal as they can."

On March 28, I sent a short email to the Letters group.

"Hi everybody, Things aren't good here. Margo is in an agony of pain in her back and hips. The technicians at the radiation unit told us the pain would most likely get worse before it gets better and that seems to be true. She's taking as many morphine pills as her prescription allows and she's still crying with pain. The morphine also suppresses her appetite and that's real bad. I feel so helpless because I can't do anything except be here to try and comfort her. I wouldn't wish this on my worst enemy and here it is happening to the one I love most. My prayers these days are for strength to get through this ordeal. Love, Dan"

Here is another entry from my diary letter.

"March 30, 6:53 AM

Mum's condition continues to get worse but she sleeps fairly well at night so that is a blessing. The other night we played eight games of Sequence before her back got too sore to sit up. She spends so much time lying down, I fear she will get pneumonia but the pain is just too much for her. She says at times it feels like her legs from the knees to the waist are going to explode.

Yesterday at the radiation unit, I talked to one of the technicians about why the last treatment made Mum so sick. She told me the radiation waves pass through the stomach area and that is why it had that effect. If it was a bit lower, the result would be diarrhea. I asked her about the increasing pain and she said it will continue to get worse with every treatment because the radiation is causing temporary swelling. By the end of next week, she'll start to get some relief. I truly hope so because she's going to be out of her head with agony. It's a good thing she has the morphine pills to help her sleep.

After the treatment, we headed to Kay's house where Mum would spend the next two nights. The sidewalk was a sheet of ice and I was afraid Mum was going to fall but she made it to the house with my support. I immediately grabbed a shovel and cleared a new path right down to the grass so they'll have no trouble getting Margo out to the taxi today. I stayed for a cup of tea and then headed home to tend the farm.

I had supper at Ruth's before heading to the hospital to sing up a storm with Bill and Joanne. Everyone was so happy to have us and thanked us profusely. There were lots of different faces this time as a lot of the other regulars had gone home. Arnold Kirby told me he's getting ready to go to the other side of the mountain and meet his maker. I must go have a quiet visit with him before it's too late. It might help him to hear some of the miracles I've witnessed in my life and he probably has a few of his own to share with me.

I picked up Mum at Kay's house yesterday morning. Mum had been using her anti-nausea pills and made it through the last two treatments okay. I had lunch but Mum didn't eat because she had her last treatment to attend.

At the clinic Mum was in the room a long time. Finally a nurse came to tell me Mum was throwing up. The nurses there are wonderfully caring and kind. They put us in a room where Margo was given a shot in her arm to stop her vomiting. It took an hour before she felt well enough to travel. She was so happy to get back in her own home with her big, clumsy man servant waiting on her hand and foot. She'd been treated well in the city but there is no place like home, especially when you're sick.

Mum settled down okay and even managed a couple slices of toast later. We both had a good sleep but she's been throwing up again this morning. The nurse told us this could persist for several days. I'm so relieved she doesn't have to face four more weeks of treatments like some people do."

On April 2, I sent a short email to Margie to answer some of her concerns.

"Good morning Margie,

Margo is going through a very rough time, feeling weak and nauseous. She's quite emotional so she doesn't want to talk to anybody yet. I'm answering all the phone calls and the machine takes the calls when I'm out of the house. This will likely pass when she starts feeling better and she'll enjoy talking to friends and family again. Right now she only wants to be with me and I remember feeling the same way when I came close to having a nervous breakdown in 1981. Oh, boy, that terrible summer seems to be a long time ago. A whole lot of good things have happened since then and I continue to count my blessings and thank God every day. We'll let you know when Margo feels up to taking calls again. Love, Dan"

The same day I sent an email to my granddaughter, Becky, to try to answer some of her questions. She seemed very worried and sending an email seemed the perfect way to answer.

"Hi Becky,

We got your fancy letter the other day and it was really nice to hear from you. It's so nice to know we have a beautiful granddaughter like you and that you are thinking about us. Gramma has been very sick from the radiation treatments she had on the cancer on her spine. She seems a bit better this morning and we hope she'll be feeling fine when you guys come to visit us this summer. I'm hoping to take you fishing at Jackfish Lake. There are lots of black flies and mosquitoes there so bring repellent with you. Did you ever cook food outside on a campfire? When you are making toast, the cook asks the others, "How do you like your toast, smoked or burnt?" It should be lots of fun and I know we'll catch some fish. Say hi to Daniel for me, please. Love Poppa"

Later I sent an email to Heather and her husband in Idaho. I repeated a lot of what I'd sent to others but here are some paragraphs that covered another topic.

"Hi Heather and Lynn,

Thank you so much for your encouraging words in your letters that arrived last week. They couldn't have come at a more opportune time as Margo has had a very rough week, the worst one yet.

This seems to have been a turning point for her and she has given up the battle. She's been incredibly strong and brave and always figured she was going to beat this cancer. She's cried more in the past week than she ever cried in her whole life and it's been very rough on both of us. They told us the nausea would fade in about four days but this, day four, has been the worst day yet. She's been throwing up all day so can't keep her morphine pills down either. Her pain must be terrible. The radiation treatments cause swelling so the pain is much worse than it was before she started treatments. I was going to take her to the hospital in town but she took a Gravol pill and it seems to have settled her stomach a bit. Now if she can only get a good night's sleep, I think things will be better tomorrow.

We both know this is going to be the end for her. The worst part is that she doesn't believe in God or eternal life and that makes me so sad. I've been trying to convince her through my faith but she says, "You can believe what you want. I don't believe any of it." I wouldn't mind so much losing her if I knew she was going to be waiting for me when the time comes for me to pass on. She says she's not scared to die but isn't ready to leave here yet. That has made her extremely bitter lately. She's reached the point where she doesn't want to talk to anyone except me and I've been taking all the incoming calls. I'm hoping for more good days ahead after the radiation does its work and hope we can enjoy the spring that is arriving now."

The next morning I sent a Margo Update email to my Letters group to let them know things had gotten much worse.

"Good morning everyone,

Yesterday was the worst day yet. She was sick so often she couldn't keep the anti-nausea pills or the morphine down and became more agitated as time went on. Steven spent a couple hours with her in the evening and that helped. He must be our guardian angel because he did the same thing with me last year when I was out of my mind with pain in the Victoria Hospital. After he left, Margo got sick again and I decided enough was enough and called our family doctor at her home. Elzette scolded me for not calling sooner and told me to take Margo straight to Eriksdale hospital.

Half an hour later the nurse was hooking her up to an IV line which would give her morphine and an anti-nausea drug. Her blood pressure was reading 150 over 91 so she must've been really stressed out and dehydrated. It usually reads about 120 over 70, just like a teenager. I'm glad she's in there where she'll be looked after well. In a couple of days she should be home.

I'm so thankful to have God on my side to carry me over the rough spots. It is said He never gives us more than we can handle at any one time. Maybe yesterday was another step to prepare us for what lies ahead. It's wonderful to have such a great support group of friends and family. I saw a lady at the radiation clinic who was facing a five week course of radiation with no family to help her. I pray God is with that lady and that He guides her steps.

The sun is shining and the birds are singing again. Two new mothers are admiring their baby calves. Life is still wonderful. Love to you all and God bless you, Dan"

It's hard to describe how ashamed I felt when I realized I'd let Margo become critically ill before I finally took her for professional help. Why hadn't I called the doctor sooner? The truth is I just didn't know any better. Ever since Margo was diagnosed with cancer, we'd been like two children lost in strange woods full of boogie men. I was trying my best to do everything right but my ignorance had cost her dearly.

My cousin Maisie received that Margo Update.

"Dear Cousin and Brother-in-the-Lord!

When you say yesterday was horrible... I'm sure that is an understatement. What you are going through is so very hard on the nerves as well as the heart and mercy strings... You are doing so fine. I want to reassure you that I can tell from what you write that you are trying to do the very best for Margo in every way. It can be such a relief to realize that, "Hey, I Am Not a Professional". Thank God—He does use others... So glad Steven could be there, and the family doctor to step in. It does lift the load of caring. I think I know of what I write but I don't want to go on and on. This is the Lord's Day... I pray it will be a blessed respite for you and for Margo. Love and hugs, Maisie Alice"

"Hi Maisie,

I just got in from the barn after a busy night looking after new arrivals. I'm very blessed to have friends and family members like you who are with us through this hard time in our lives. Jim phoned yesterday and asked what they can do to help. I told him to phone once in awhile or email and to please pray for us.

My faith in God is real and is growing stronger every day. As I walked to the house just now in total darkness, I looked up at millions of brilliant stars and was overwhelmed with happiness as I thought about my countless, undeserved blessings. I leaned against a gate post and had a good early morning chat with God and so my day has started off well.

Margo is doing much better. I visited her twice yesterday and she's asking me to let her friends know she is ready for visitors. That's such a relief for me as she hasn't even talked on the phone for over a week. She's the kind of person who keeps things bottled up inside until she has to let off steam. She hates to let her emotions show and sometimes that has caused rough spots in our marriage. At times she might've been mad at me for a week while I went merrily along with my busy life without realizing something was wrong. She's been crying at the drop of a hat for a week now, after realizing things aren't going to get any better with her fight against this damn cancer.

I had some wonderful replies to my Margo Update yesterday. I printed them, including my Update, and took them in to Margo last night. I don't want to constantly pressure her with words but I think as she reads these messages, she will feel the faith that shines through. My worst fear is that she will pass on without coming to Christ. That is probably my fault because I turned away from God for years and didn't even take my kids to church. It was easy to say, "We'll let them decide for themselves when they get old enough." That is what Margo's parents did, so she has no knowledge of God's word like I was fortunate enough to have had as a child. They did have her baptized but that's as far as it went.

I am rambling on here. Have a great day. Dan"

The phone was also ringing non-stop. One call was from Margo's cousin, Shirley Nordal. We talked for a long time about our faith in God and how He can carry us through bad times like this. Shirley and her sister, Darlene, have extremely strong faith and it helped strengthen my own faith when I talked to them.

The following day, Tuesday, April 5, I started another journal letter.

"Yesterday Margo had so much company the nurses finally put a sign on her door allowing only two visitors at a time and then only for ten minutes. I'd been to see her in the afternoon and it was standing room only. It tired her out but she was back to loving the attention and feeling blessed with so many good friends.

Later in the evening we had a good visit alone. I feel she's going to bounce back from this latest setback and if the radiation does what they claimed, she'll have less pain to deal with. I've been getting kind of strung out lately with all that's going on and have been having a hard time sleeping at night. I fall asleep okay but after doing the two o'clock pen check, my mind goes into high gear. It's impossible to get back to sleep until it's almost time to get up. I must look after myself because I won't be much help to Margo if I go off the deep end."

The next day I sent this email.

"Hi Tracy,

Mum is still in the hospital. She is weak and her stomach can't hold any food down yet. That radiation really must've burned her poor stomach.

This morning she walked with me all the way down the hall to little Annie Pottinger's room in the care home. After a short visit she walked all the way back. I don't know why the nurses haven't had her up more. She's reluctant to get moving but I tell her she doesn't want to end up with pneumonia or lose the use of her legs.

Please try not to worry about her, as she is in good hands. Love, Dad"

Thursday started out well but got totally out of hand that evening, as I wrote in an email to my Letters group.

"Hello everyone,

They've been keeping Margo on intravenous to keep her fluids up. She's also been getting a steroid, to heal her stomach that was burned from the radiation. She receives morphine for pain and an anti-nausea drug. She's starting to eat semi-solid food but is having trouble keeping it down.

This evening about five o'clock I was ready to have supper when a nurse phoned. Margo had been acting in a strange manner. She'd climbed out of bed, put on her jeans and blouse and was walking down the hall with her IV pole. She said she was going home. When the nurse told her I wasn't there yet, she decided to go through to the care home and visit Annie like we did yesterday. The nurse walked with her and noticed she was acting rational one minute and totally strange the next. I told her I'd be there right away and ran out of the house without even locking the door.

I cried all the way to town and it reminded me of the time I left town after getting the call that my mum was critically injured—I cried all the way home that day. This time I was going in a different direction and I wasn't trying to make any bargains with God. The only thing I asked was that He would make me strong so I could quit crying before I saw Margo. I became calm at once and was able to keep a smile on my face when I saw her. A technician was trying to get a needle in her shrunken veins so they could check her blood and Margo seemed very tired. Half an hour later they had the test results and all looked normal.

The nurse phoned Dr. Steyn and was told to take her off the steroids. Five days on that medication could be causing these strange symptoms and I'm praying things will be better tomorrow. My fear is that the cancer has reached her brain and if so, it will be all downhill from here.

I sat with her all evening while she slept. I quietly stood up once to shut off the fan and my knees cracked. She was wide awake at once asking if it was time to go home. Several times during the evening she asked about going home and once she asked if we were leaving for the city in the morning. She asked all sorts of questions that made perfect sense, like had I checked the sump pump in the basement to make sure it was working?

Had I picked up the beef supplement in the morning? The next minute she would come up with something right out of left field. It was hard to keep her from seeing my tears. I don't mind crying but I didn't want to agitate her.

It was strange as she kept drifting from being perfectly rational one minute to not making sense at all. One time she woke up and mentioned that I looked really tired. I told her I was having a hard time staying awake. She told me to stretch out on the bed with her and sleep for awhile. If I hadn't been so concerned about her, it would've been funny because the bed was a typical, narrow hospital bed and there surely was no room for the two of us. I tried to tell her that but she insisted there was plenty of room. I guess maybe there was but I didn't know what the nurse would say if she came in and caught us cuddled up together.

The nurse agreed to move her to a private room near the desk where they can watch her better. The other room was at the end of the hall and around the corner. I might sleep easier knowing she is well looked after.

Let's pray for her tonight. I'm so thankful for my loving family and will be depending on you in the future. Love, Dan"

"Saturday morning Margo Update April 9, 2005

Hi guys,

I was at the hospital early yesterday morning and found Margo looking a lot better, with the confusion pretty well gone. The steroids must've been causing the confusion. The day nurse said when she came on shift at seven o'clock Margo was emotional and weepy. It turned out she hadn't had morphine since 3 PM the day before, causing pain and withdrawal symptoms. I don't know what happened there—the doctor had ordered the steroids be stopped but I didn't hear anything about cutting back on morphine.

Margo had been running to the bathroom all night and was tired. I noticed the night before that her ankles and lower legs were quite swollen and once she was off the steroids, which cause water retention, she had to get up to pee every few hours. I spent several hours with her and she was so tired I couldn't convince her to get up and walk with me.

She had a fair amount of visitors and was happy to have them so that was great. Dr. Steyn says Margo could be in there for quite awhile yet. I'm so happy the hospital isn't crowded so they aren't rushing her to go home. From now on she'll be on a lower dose of morphine every three hours to keep her relatively pain free.

We played one game of Sequence but she was so tired she couldn't concentrate and was missing moves. I swore I'd never do it but I ended up throwing the game. I had the cards I needed to end it but waited till my sweetie pulled the two-eyed jack she needed to beat me. I couldn't stand the thought of beating her on this game. She was settling in for a good sleep when I left. Love, Dan"

For the past two years we'd noticed that some of our friends were able to stay close to us while others pulled away. We'd been there and done the same thing when friends of ours were dying and we knew how they felt. It surprised me to get an email of support from my niece, Karen, and to feel the level of maturity from such a young woman. Here is my reply.

"Hi Karen,

Keep praying for us. I admire you for having the courage to send kind words of support and love. So often we shy away from someone who is terminally ill because we don't know the right words to say. We have found by our own travels through this horrible cancer ordeal that there really aren't any RIGHT words. A simple message of support like you sent is all that's required. I'm hoping to see you guys again some day in happier times but it will be strange not to have Margo with me. Love Uncle Dan"

"Sunday morning Margo Update April 10, 2005

I was so happy last night to find Margo much improved. She was well rested and feeling stronger and readily agreed to walk down the hall to visit Arnold Kirby and his girlfriend, Sandy. Margo had eaten two little cups of yogurt for supper but was getting sick of the bland diet. We asked for toast and she ate it all with another yogurt. She had quite a few visitors but they arrived at different intervals so it didn't tire her out. I was afraid she was never coming home but now feel much more confident. I actually slept all night. Dan"

I was still making time to keep up with my journal letter.

"Margo had more company in the afternoon, including her brother, Victor, plus Kay and her husband, Alf. Donna Smith was up to see her too. It must be hard for Donna to visit after what she went through with Lorne and I sure give her credit for being strong enough to do that.

April 11, 8:45 AM

Yesterday was Sunday and I decided to go to town a bit later and then go to church in the afternoon. I spent the morning feeding cattle, bedding pens, doing laundry, making phone calls and sending emails.

Margo seemed much improved and told me she'd eaten a good breakfast and a nice chicken lunch. While we were visiting, Bubby Reykdal dropped in. She is a special friend of Tracy's and we've gotten to know and love her over the years. She lives way out near the lake up near the Narrows so it was sure nice of her to make the effort to come in to town. She brought Margo a tin with matrimonial cake and it was mighty tasty.

At quarter to two I took off to church. John Ives did the sermon and it was plain to see he had done a lot of work to get it ready. After the service I stayed for only a few minutes of fellowship before heading back to the hospital. Ruth was visiting and helping Margo who had been losing her fine lunch. I guess it's going to take time for her stomach to heal properly."

The next day I emailed Donna Smith to thank her for courage.

"Hi Donna,

I want to let you know how much it meant to Margo and I to have you visit her yesterday. I admire your strength and know it must've been very hard for you to be there. Margo is much stronger today and I think it won't be long before I can bring her home again. We have a wonderful support group and with God on our side we'll get through this ordeal. I'm so thankful for every day I have her. Dan"

I was noticing people who had been through this same sort of ordeal seemed to be incredibly strong. This realization helped me at this point because there were times when I wondered if I would be strong enough to be there for Margo when she needed me most. Here's another entry from my journal letter.

"April 12, 4:25 PM

Yesterday morning I found Margo feeling strong and happy. She told me she wasn't getting any morphine through the IV line, just the same pills as before and she hadn't needed one for many hours. She had a visit from Emile Lavallee and his wife, Marcelle, from St. Laurent, who brought her a huge bouquet of flowers. Emile was the fiddle player in the band that played for our anniversary dance three years ago.

Last week, I got a call from Dr. Quack wanting to know how Margo was doing and why we hadn't been back in to see him. I told him straight out I wasn't feeling very happy with him. I let him have both barrels as I tore a strip up one side of his hairy hide and down the other. He was shocked to hear I'd gotten the medicine for $200 per bottle when he'd been charging us $1250. He immediately put the blame on his broker in Montreal and said he knew the guy had been ripping him off. I told him someone has been ripping me off and I'm suspecting it's him.

I told him that if I, as a beginner Internet user, could find this information, he could've done the same. He said he is a dinosaur with computers and is up to his butt in work so had no time to check it out. I told him he has a lot of desperate people who can't afford to pay those prices and I think he could've taken some time. He thinks nothing of sitting for two hours straight giving you his line of BS until he has you firmly hooked, with your cheque book in hand. He told me he would call his broker and call me back.

A few days ago, he left a message saying he'd been telling his broker off and the guy hung up on him. He also found out from his accountant he's going to have to pay GST on all those extra charges. Well, boo hoo for him. I won't believe a word he says from now on even though he ended the message by assuring me sincerely he isn't an unscrupulous crook."

Our support group was growing larger every day as I added new members to my address book and they forwarded the Margo Updates to their immediate families. It was incredible to receive so many messages of support.

"Tuesday Margo Update April 12, 2005

Hello everybody,

After a tough two weeks Margo seems to be recovering nicely. Our prayers have been answered and she is on the mend. The old sparkle is back in her eyes and that's good to see. She's going to ask her doctor this afternoon for a pass to go to the dance on Saturday night. It's only a few blocks away, so if she gets tired or weak, I can get her back to the hospital quickly. A couple days ago, I printed one of Margie's messages of love and support and took it in for Margo to read. She cries every time she reads it but she loves it so much she reads it over and over again. This morning I printed out a whole bunch of loving messages from friends and family from near and far and put them in a binder. She spent many happy hours reading them so I'll be doing that again as the messages arrive.

We talked for years about hooking up to the Internet but for some reason or the other we never felt the time was right. This past winter we took the plunge and now these messages are proving that the timing was exactly right—another example of the way we follow the Master's plan even when we don't realize we are doing it. Love, Dan"

Here's another journal entry.

"April 14, 4:45 PM

Yesterday morning I found Margo feeling stronger again. We took the Sequence board to the TV room and tried to play but there were constant interruptions with people coming by wanting to know what we were playing. Some people knew the game but most hadn't seen it. A crew of men was busy tearing out the old floors in the hallways so all the incoming foot traffic was being diverted past where we were sitting.

I left about noon and had barely gotten home when Margo phoned to say her stomach is still rebelling and she'd thrown up her breakfast. I was on the road back to town by four-thirty. Margo was feeling well again and ate a good supper while I was there. Bill and Joanne came along and Joanne sat with Margo while Bill and I went singing in the hospital and care home.

April 16, 8:35 AM

On Thursday morning I spent time with Margo at the hospital and then Ruth and I went to Moosehorn to pick up our beef. On the way home I picked up Margo and brought her home for a visit. We had a peaceful afternoon but when I was out doing chores Margo got sick and threw up. Later she ate some supper but threw that up as well. We watched television together for awhile before I took her back to the hospital.

In spite of her being sick, it had been fun having her home. She loved being out in the bright sunshine, marvelling at how the snow had disappeared. I opened the window in the screen door and we covered up with blankets on our big couch. We both fell asleep listening to the wind in the trees and the birds singing. It was such a wonderful moment when we both woke up at the same time and were able to share a happy smile with each other. I'm looking forward to having her here again soon if we can just get her nausea under control. Love, Dan"

The next day I received an email from Maisie. She suggested we call the binder that held all of the emails, Margo's Happiness Binder. That name really stuck and from that day, every email that came in or went out was put straight into that special binder for Margo to read over and over again.

ELEVEN
Margo's Happiness Binder

April 16, 2005 began with brilliant spring sunshine and blue skies. I was on top of the world as I went about my chores and cleaned up around the yard. There was a very good reason for feeling so happy. Elzette had given me permission to take Margo out for the evening to the dance in town. Only two weeks earlier Margo had been so sick it was as if she was getting ready to leave me at any minute. Now here she was bouncing back and eager to go dancing.

It was strange but wonderful how our support group continued to grow. Tracy was now sending my emails to her co-workers. One of her friends, Anita, asked for and received permission to send us an email. It was amazing that people we'd never met were praying for us and that they would take time from their busy lives to send beautiful messages of support.

"Dear Margo & Dan,

I'm very glad to hear Margo is feeling better. I felt compelled to drop you a note to let you know there are many people who you don't even know that keep you both in their thoughts and prayers, and are sending well wishes your way. I wanted to let you know how much Tracy's co-workers at Faneuil care about her. I've worked with Tracy for ten years and recently she was put next to me at work with our new seating plan. She is such a pleasure to sit beside. No matter how bad a day she is having she never takes it out on others and is never a "downer". My daughter, Christine, is Tracy's age and they became friends when Christine worked at Faneuil. When I had cancer five years ago, my daughter was my "rock" and I know that Tracy is yours. Your daughter is a reflection of what great parents you are because wonderful people are only created by wonderful parents. I will be thinking of you. Best wishes, Anita Labossiere"

"Hi Anita,

You are so kind to take the time to write. I would hate going through our present troubles without such a wonderful support group praying for us and wishing us well. My prayers these days are to have Margo as pain free as possible and have her with me for as much precious time as there is in the Master's plan.

Margo is rallying now. I'm taking her to the supper and dance in town tonight. There's a chance she might throw up while we're there but if that happens we'll deal with it the same way we deal with her illness—straight on and chin up. She's looking forward to going out and seeing her friends.

I'll put your message in Margo's Happiness Binder. She treasures every one and reads them over and over. Love to you and your family, Dan"

That evening turned out to be more wonderful than we hoped. I was up early the next morning sending an email.

"Hello everybody,

It's a beautiful morning and great to be alive. Margo loved being at the dance last night, seeing her friends, and she even managed a couple of turns around the dance floor with me in an old-time waltz. Our fiddle playing friend, Emile, saw her struggling and kept the tunes short so she didn't have to sit down before the song ended.

Hey, everybody, how about sending her messages by email? I print these wonderful messages and put them in her binder to read. She gets great pleasure from them and spends hours reading them. She still tells me she doesn't believe in God but somehow I think she's starting to believe our friends can't be wrong about such a wonderful thing.

Bill asked me to play guitar and sing a gospel song with him in his church in Ashern this morning. We're going to sing one of Elvis Presley's gospel songs called "Who Am I." Every time I sing that marvellous song, I ask myself that same question. God does love us all, every last one of us. I hope you are well and happy on this great day. Love, Dan"

A couple hours later, the replies started coming in. Here's a special one from my grandson, Daniel, who was a computer whiz already. He asked me a question that must've been weighing heavily on his mind. I carefully considered my reply.

"I really feel bad for Gramma because she is really sick. Is it true Gramma is handicapped? My mom tells me she is well. Goodbye for now. Daniel

P.S. Say hi to Gramma for me."

"Hi Daniel,

I was happy to get another email from you. You asked me a very hard question about whether Gramma is handicapped. She isn't handicapped like you are when you can't move your arms or legs but she is so sick; I guess you could say she is handicapped a little bit. She can't do all the things she likes to do but she is still happy and doing fairly well. This cancer she has is a very cruel thing because she isn't ever going to get better and recover from the disease. We just have to love her and hope she'll be with us for a long time yet. I'm hoping she will come to believe in God so she will go to Heaven and be safe there after she dies. I know this is very hard for you to understand but you are getting big now, almost ten years old, and I think you are old enough to know the truth. You have your Gramma's blood flowing through your veins and I know you are very brave, just like she is. I think she is the bravest person I know and I'm so lucky to have her for my wife. She gave me four fine children and they gave us eight beautiful grandchildren. We are so proud of all of you. Remember that both of us love you and Becky and we want you to grow up to be good people. Love, Poppa"

I shed a lot of tears that night trying to answer Dan's question but those tears didn't compare to the deluge that occurred when I read this message from Donna Smith the next morning.

"Dear Dan and Margo,

I think it would be a miss not to tell everyone out there how beautiful Margo looked last night at the 200 club dinner and dance. The first thing one noticed was a warm and lovely smile and eyes that lit up each time she greeted an old friend. And believe me there was a line-up at her table of friends waiting for a chance to say hello to this very brave and courageous lady.

Her hair has grown out to be her natural color, dark brown with no gray in sight. Her friend and hairdresser, Lorraine came to the hospital, wheeled her to the chronic care wing and washed and styled her hair which complimented her angelic face. Her daughter-in-law, Linda, groomed and coloured her finger and toe nails in a lovely soft colour and she looked elegant and surreal.

The memory of Dan and Margo dancing is one I'll always treasure and I hope you'll all enjoy hearing how brave and wonderful these two people are and know that they are surrounded with love and good wishes.

Margo may or may not believe in God and that's ok in my opinion, because we are all at different levels of faith or lack of it for reasons unknown to us. I believe the spirit of the Lord is at work in Margo and Dan and this spirit is there to help them to complete their life's journey.

Margo, you touch all our hearts as we watch you do your best and do it in style. We love and admire the lessons you are teaching us all. May blessing flow on your paths now and forever. Love, your friend, Donna Smith"

"Good morning Donna,

Wow, God must have been working in you in the wee early hours today. What a beautiful message and coming from you, after all you've been through makes it even more special. Our memory of Lorne remains as strong today as it was when he left us. After years of hunting together, I still feel Lorne's presence with me as I crouch in my blind listening to incoming flights of geese while the first hint of dawn lights up this beautiful land we live in.

Wonderful family and friends like you have been carrying us through this rough journey and making it easier to bear. Thank you for taking the time to think of us. Love, Dan"

Somehow I was still finding time to churn out my journal letter in great detail and here are a few paragraphs gathered from the happenings of April 18.

"I finally got away from here but while driving out the lane I noticed a cow stretched flat out in heavy labour. I went back in the house and phoned the hospital to let Margo know I'd be late. It only took about twenty minutes before the new calf flopped out onto dry grass.

I found Margo looking quite well but she said she'd come close to throwing up in the afternoon. That has been the pattern with her getting sick in the afternoon every second day. Today seemed to be a break in that pattern and she never did get sick.

We spent a pleasant hour together playing Sequence. The cards were in her favour and she beat me three games to two before we quit. Then we did some walking back and forth in the halls. She is walking much better now and certainly doesn't seem to have the pain she had before. Maybe this treatment did work the way it was intended; it's too bad it had to burn her stomach at the same time. When she got tired I tucked her into bed with her Happiness Binder."

On Wednesday evening, April 20, I sent this update to our support group.

"Hi guys,

I picked up Margo at the hospital this morning and we headed for the city. First stop was at our car dealers where we had the oil and filter changed. Then we drove to Kay's house and had a bite of lunch. Margo rested before we went to the Cancer Clinic about one o'clock. We barely got in there when she started feeling sick. The nurse took us into one of the examining rooms where we had some privacy and not a moment too soon. Margo's spaghetti lunch came up and while I was holding the pan for her, I almost chucked my lunch as well. That would've been a disaster because I'd eaten about five times as much as she had. As soon as she felt a bit better, we started giggling about how funny it would've been if the doctor had walked in just in time to have me barf on her shoes. Somehow we seem to find humour even at the worst of times and that's what keeps us going.

A few minutes later Dr. Vijay came in and asked a few questions. We suspected this appointment had been scheduled so the doctor could do a follow-up report and we were right.

The trip home was okay and Margo came out to the farm with me for about four hours. She had some of the nice supper I made and it stayed down alright. I took her back to the hospital about eight o'clock and got her tucked in. I was still there when Dr. Steyn dropped in to check on her. She had a plan in mind for us so I'm glad I was still around to hear it. She can arrange for us to get as much home care as we need, up to 24 hours a day. Margo has been enjoying all the visitors she gets in town and the wonderful care from her nurses so at first she was kind of lukewarm to the idea. Elzette told her she is much too well to be lying around in there where she's in danger of picking up some of these nasty viruses that are going around. She can arrange for any supplies we need, like a walker and a wheelchair. I'll see about installing a wheelchair ramp and we'll keep her here with me as long as possible. We think we'll go with help ten hours a day and I should be able to handle the rest.

I'm so thankful to Elzette for taking time to inform us of the possibilities. She was on the phone five minutes later talking to someone about it. Tomorrow she'll have the head nurse go right to work on it. The home care workers will do cooking and look after Margo by bathing her and washing her hair but they don't do any housecleaning. I think it's kind of stupid to have an able-bodied person hanging around all day and they can't clean house, nor do laundry. I'm not complaining, mind you, and am very grateful for any help. I'll be able to do my farm work without leaving her alone in the house and she'll be right here when I come in. It's been a long day and I'm totally wiped so it's off to bed to sleep the night away. Dan"

Journal letter,

"April 22

I woke up to heavy frost yesterday morning, the first in weeks it seems. That may slow down the trees that are on the verge of breaking out in leaf.

I'd been invited to an appreciation tea at the care home at two o'clock so went in early to visit Margo. She was having a rough day and had been kind of sick all morning. She told me Arnold Kirby, in the palliative care room, was in pretty bad shape but he'd been up walking the halls first thing in the morning as usual.

Margo phoned me before bedtime to say that Arnold's family had been called in because he wasn't doing well. I was tired and fell asleep at ten. Shortly after 11:30 PM, I woke up suddenly, feeling that Arnold was there with me. I sensed that he'd passed on and it took me over an hour to go back to sleep. This morning Margo told me that Arnold died about four this morning. I'm not putting any great significance on this as I often dream about something we talked about the day before. I do know that Arnold is out of his pain at last, after several years bravely fighting stomach cancer. He told me a few weeks ago that he was getting ready to go to the other side of the mountain. I wish my old friend well and hope he finds lots of big bucks and fat pickerel there."

That evening I wrote a happier note in the form of an email with the subject line, "She's home again."

"Hi everyone,

She's here and doing fine with the help of a nice home care worker. Our doctor got things rolling in a hurry once we gave our approval. We're going to get help ten hours per day during the week and three hours per weekend day. Margo is so happy to be home where she can look out her window and see our little calves running around the front pasture.

Tracy and The Boy are on their way out to spend the weekend with us. We haven't seen Dylan since February. He is doing quite a bit of talking now so we're going to see a big change in him. Dan"

Here is a paragraph from my April 24 diary letter.

"I got a call from Arnold's daughter, Sandra. Bill and I have been singing every week for Arnold and he'd told his family he wanted us to sing at his funeral. He also wanted Bill to do the service but unfortunately, Bill won't be here. I told her she could count on me to sing. I confided to Sandra that I'd been quite sure in the night that her dad had died and felt he was here with me. She said the same thing had happened to her Uncle Norman."

On April 26 Margo received an email from Linda Nickel that brought back a lot of memories for us.

"Hi Margo,

You've been in our thoughts a lot lately. One of the things I was remembering about you was our first visit to your home. You still lived in the old, log house at the time and Dan was overhauling a motor in the living room, one of the bed rooms had a batch of baby chicks in it and you had to duck under plumbing pipes.

You served us a beautiful roast beef dinner, and your home could've been a castle, as you made us feel so very welcome, and you could feel the love that was there between you and Danny.

I've always remembered that visit through the years, and whenever I became dissatisfied with my home I'd remind myself how you made me see how it didn't matter where you lived or what your house looked like, having your family together was what was important. You and Dan have been good friends and an inspiration to us through the years.

I'm glad to hear you're getting some home care to help out with things. Dan is a wonderful help, but he has the farm to keep going too. After our last visit I asked Elmer if he could've made such a nice meal like Dan had done, and he thought it might've been peanut butter sandwiches. Love, Elmer and Linda"

"Hi Linda,

Thank you for your kind words and fond memories. Margo was so touched by what you said and asked me to answer you right away.

You're lucky you weren't at the party in our old house the time I was thawing out a frozen coyote in a corner of the kitchen. As he thawed, his belly started to swell and he kept letting out sneaky little farts that just about gagged the lot of us. Margo blamed me for doing it and finally, she came charging out of the bathroom with a can of deodorant spray and gave it to me full force on the back side of my blue jeans. We finally tracked down the real culprit and threw it out of the house. Of course I dragged him back in the next day to thaw and skin him. How many young wives these days would put up with mischief like that? Margo has been a one in a million gal and I'm so lucky to have her. Dan"

Here's a Margo and Dan Update April 28 2005

"Good morning everybody,

Margo has her fighting spirit back and she proves that to me over and over as we play our favourite Sequence game. That simple game has been a blessing to us over the past two years. I'm happy to say she hasn't thrown up since she's been home and that will be a week tomorrow. It's quiet and peaceful here and she has plenty of time to rest or talk to her workers or me as she chooses. Our workers are angels in disguise

Yesterday morning dawned cloudy with a cold wind gathering moisture off the open waters of the lake and dumping it back on us in the form of snow granules and ice pellets. At six AM I found a new calf shivering in that icy wind, too cold to make the move to survive. I got her on her feet and a few minutes later she was standing at Momma's Snack Bar having her first real meal.

We had an appointment with Dr. Steyn and left early so we could drop around by Lundar to pick up an order from Sears. Margo had done some shopping by phone and blew a good portion of the contents of her last brown envelope. True to her generous nature, one of the most expensive items was a pair of dancing shoes for me to replace my worn-out ones.

We discussed Margo's condition and care with Elzette and then I spoke to her about something that had been bothering me. For the past year, I've been having bouts of pain in my gut but convinced myself each time that it was muscle strain or indigestion. The pain has been getting worse and feels like the pain I had with the blood clot in my liver about seventeen months ago.

I hadn't had breakfast yet so she phoned the hospital and arranged for blood tests to be done in early afternoon. I took Margo home in time to meet her care worker and a therapist at noon and then headed back in to town. When I stopped at the clinic to pick up my requisitions, Elzette told me she'd booked an appointment with Dr. Lipschitz, my liver specialist, for May 2. Once again, I'm in awe of our doctor for the way she can get things done in a hurry.

We planned on leaving about six-thirty for our jam session in town but one of our heifers was calving so we waited an hour until she calved. We had a wonderful time at our party. The music and singing was marvellous. Love, Dan"

April 29 was rather rough. The time had come to get our house in order, as Margo's oncologist had suggested. Here is what I recorded in my journal.

"We went by Ruth's place so she could witness Margo's signature. The lady in charge of health care in this area suggested we fill out and sign a form for Anticipated Death at Home Protocol. This means if Margo should die at home, we can just notify the funeral home and we won't have the house swarming with cops trying to investigate a crime scene.

I dropped the form off at Dr. Steyn's office. She'd already filled out her part of the form. She will see that a copy is distributed to the Coroner, RCMP, some other official in health care and the funeral home. It's strange how a year ago we couldn't have dealt with anything like this but one crisis after another kind of prepares you for it. We are thankful we have people who are trained and caring enough to help us. We are also thankful for Dr. Steyn, who is making sure that Margo won't suffer at all, if possible. We were able to talk to her the other day about the way we want things done when Margo lands in hospital for the last time, if in fact it comes to that. We don't want heroic measures or feeding tubes that prolong the inevitable but we do want her to have intravenous fluid so she won't be dehydrated and uncomfortable.

By seven o'clock, we had company, Hulda and Albert and Raymond and Georgina. These were the first guests we've had over since Margo got home and it was quite appropriate because they were the same two couples that came straight here the day after we found out that Margo had cancer. We visited for an hour and then pigged out on a beautiful trifle that Ruth brought me as a pre-birthday cake.

After Ruth left, we played six games of Sequence. It was frustrating for the gals because the guys kept getting all the good cards and we beat them five to one. They are good sports in spite of being fierce competitors and they never got mad at us. Later we moved to the living room and listened to Georgina reading some of the beautiful messages of support in Margo's Happiness Binder. They all left about eleven o'clock and I tucked Margo into bed. She fell asleep without using a sleeping pill and slept until almost eight this morning. She hasn't had any morphine for the past two days except for the long-acting pill that she takes twice in 24 hours, so things are looking up. The extra morphine which is taken as needed for pain makes her a bit confused so we are happy that she's able to do with less. She's back to looking after bills and writing cheques again and doing a good job of it."

Here is an entry from May 1 regarding an amusing incident that happened the previous evening.

"What with one thing or another it was getting a bit late by the time I got the afternoon chores done and we started getting ready to head to the supper and variety show I was taking part in. Margo had taken a morphine pill earlier for the pain in her back and the pill had her kind of keyed up and excited. When that happens, she finds all sorts of little chores for me to do. Usually she wants them done "right now" to the point where I'm forced to leave one job to start another. It used to make me impatient with her until a friend explained that when you're on drugs, you tend to live for the moment. I know it's the drugs doing this so I try my best to keep my mouth shut and get the chores done as quickly as possible. She tells me that if I don't do them right now, I might forget them. She has a good point there.

I wanted to do a bunch of shopping before we went to the hall but by fulfilling Margo's requests, time was rapidly running out. I knew it was my fault for not getting out doing chores sooner so just kept on plugging away. One of the jobs was rearranging the bedroom so we could change sides on the bed. The therapist suggested the switch to make it easier for Margo to get in and out of the room. After over forty years of having our own sides of the bed, we switched sides.

When I finally had things all arranged, I dashed into the bathroom and stripped naked. I'd barely finished shaving when Margo said the vacuuming hadn't been done and the rug was a mess. I grabbed the vacuum and went at it, doing a very thorough job. It could've got by with a lick and a promise, as my mum used to say, but I guess the perfectionist (or the devil) in me made me take my own sweet time and do a really good job. If Margo was impressed by having her very own man servant vacuuming the floor in the nude, she didn't say, but she seemed to like the way the rug looked when it was done."

Another Margo and Dan Update

"Hi Folks,

Things are going pretty well. Margo is about the same, kind of weak but still hanging in here. We went to the supper and benefit variety show on Saturday night and Margo managed about five straight hours sitting on a hard chair. Donna Smith kindly brought in her soft booster seat from her car and that really saved the day. I was one of many performers singing and playing and it was a fun evening as we raised about $2300 toward the purchase of an ultrasound machine for the hospital. Margo wanted to make a good donation toward the hospital so she spent quite a bit of loot on tickets for the silent auction. Her generosity paid off when she won eight of the fifty-two prizes.

I was at the Health Science Center this morning to see Dr. Lipschitz. He couldn't find anything abnormal except for some tenderness in the region when he palpated my liver. He said it's quite possible there isn't a normal flow of blood to my intestines so he ordered a CT scan and a gastroscope. It sounds like two more trips down that long highway in the near future but I'm relieved to be finally doing something about it. Dan"

On May 4 which is my birthday I drove to the city to see Dr. Lipschitz again. He performed a gastroscope on me which showed an ulcer in the duodenum just below my stomach. He said the ulcer was getting pretty big and quite deep, almost to the point where it was going to start bleeding. I told him I never thought I'd be happy to hear I had an ulcer but I was just so thankful the blood clot wasn't back in my liver. He wrote me a prescription for medication and said I'd most likely have to stay on it the rest of my life.

On May 9 I sent an email with the subject line, "Update for our kids". Margo wasn't doing well and I wanted them to realize she might not be with us much longer.

"Hi Kids,

I can see Mum going downhill fast these days. The morphine makes her constipated and it's hard to figure out how much laxative to give her. The laxative makes her stomach hurt so it's all a balancing act. She isn't eating much and is losing weight. We don't know how much because she is too shaky to get on the scale. Last night she ate a cup of stewed prunes and later threw them all up right after taking her long-acting morphine pill that she gets every twelve hours. After she settled down, I gave her a short-acting morphine pill and half a sleeping pill and she slept through the night. Half an hour ago she woke up in terrible pain. I gave her another short-acting pill to tide her over until eight o'clock and I think she's gone back to sleep.

She told me she wished she was dead. That is the first time I've heard her say that and it broke my heart. Her pain seems to be getting worse every day and now it is in her neck as well. I told her that when she's ready to go, she can quit eating and we'll just deal with the pain. We are lucky we can talk freely about this but we really don't have much choice in the matter.

This is such a cruel disease and to see it taking away someone I've loved since I was sixteen years old has to be the hardest thing I've ever faced. I don't know how I'm going to look after her when she becomes too weak to walk. Her bones are so painful; I can barely touch her anywhere. How am I going to lift her in and out of the wheelchair if the pain is too much for her to bear? I've built up the area west of the house with clay and crushed rock and Ben is going to help me build a wheelchair ramp there this week. Hopefully we can have more good days together yet before she has to go into the hospital for the last time.

I know all of you are hurting as much as I am, maybe even more, and there is nothing we can do except try to make her as happy as possible. Your phone calls mean so much to us, so keep them coming. Mum was so proud that all of you called or visited on Mother's Day yesterday. Your love and support means so much to her.

Prepare yourself for the worst. We have to be brave like Mum and face this head-on. We love you guys and are proud of each and every one of you. I can't imagine getting through this without your support. Love, Dad"

On the eleventh of May I sent this note.

"Hi Donna,

Things are about the same. Mum is hanging in here and seems to be doing okay this morning. It really froze hard last night and was minus 6 C when I got up. Ben came down yesterday and we built a wheelchair ramp. I don't know if it will get used much, as Mum just wants to stay at home where it's comfortable and quiet. Maybe when the weather warms up we'll do some driving around in the evening and enjoy nature at its finest. I love you lots, Dad"

On May 13 it was time for another Margo Update. Email was wonderful in the way it let us keep in constant touch with our loved ones.

"Hello everybody,

Three weeks have gone by since I brought Margo home from the hospital. Our home care workers are kind and caring and helpful with all of Margo's wishes. I see Margo every day but I'm still shocked at how quickly she is failing. She has no appetite but her will to live drives her to eat. She loves the peace and quiet here without even the television on most of the time. We keep the window open so she can hear her birds. Her beloved wild canaries came back a few days ago, in spite of the cold weather.

Margo's pain is terrible now. She can hardly bear my touch to bathe her or help her move around. We've switched sides on the bed and I keep rolling toward her in the night and bumping her so she wakes up. I offered to move to the other bedroom but she wants me close in case she needs me in the night.

Donna is flying back from Edmonton on Monday and will stay five days. It will be good to see her again. Steven is home from his job in Ontario and he and Linda brought out Jayden for a visit with Gramma yesterday. We all love the little guy.

My own health is much improved since I started taking medication for my ulcer. I'm feeling so much better; it's hard to believe how much pain I was in all winter. I guess I was too preoccupied to notice. Well folks, God bless you all and keep you safe. Love, Dan"

Early the next morning I sent this special email to our kids.

"Dear Kids,

Mum seems to be getting much worse very quickly and I'm not sure how much longer she's going to be with us. This morning she woke me up saying she wants to go back to the hospital. I gave her a pill and she is sleeping again. I want to keep her here as long as I can because the hospital floors are being renovated. Some of the rooms are shut down and the rest are jammed full. There are even patients on cots in the hallways. I will talk to one of the nurses this morning about bringing her back there when they get a free bed. Mum and I have talked about it and we both know that when she goes in this time, she won't ever be coming back home.

Donna, we are so happy you are coming on Monday. Both of us are quite sure she isn't going to be with us in July so it's good you are coming now.

Kevin, don't feel bad that you can't come home right now. Mum and I talked about that and we both know you have to catch those prawns while you can. You came last fall to see her and we saw you out there in January. We're so glad now that we took the trip.

Steven, thanks for coming back early. It made Mum so happy to see Jayden the other day. Her grandchildren are the light of her life and I think her biggest regret is that she won't get to see them grow up. If my faith and beliefs are right, she will still see all of us even after she's gone from her frail body. I pray with all my heart she will find faith in God and Heaven before she passes on.

Tracy, thank you for being our rock. You were there for us all through last summer and I know you'll be here when we need you again.

To all of you, thanks again for growing up to be good people that we're proud of. You all gave us bad moments at times but that is just part of spreading your wings and growing up. I was far from being a perfect parent but I believe Mum was as close as they come to that state, much like my own dear mother was.

Please excuse me for emailing instead of phoning. I'm kind of emotional right now and I find it easier to express my feelings here instead of on the phone. It's taken me a long time to type this note because I can't stop crying. I love all of you. Dad"

I was still making time to write my daily journal but was trying not to repeat what had already been written in emails.

"May 14, 2005

Margo got a call from Garry Seidler asking if they could come over and she told them to come in the evening. We visited for awhile and they were amazed when Margo got up off the couch and came into the kitchen for Sequence. We played guys against gals, and the gals won the rubber match. Margo finally had to give in and retreat to the couch and they went home.

For several days she's been getting terrible pains in her neck. The cancer must've travelled up her spine and is tormenting her there."

On May 16, I wrote a lengthy Monday morning Margo Update.

"Hello everyone,

Saturday morning Margo woke me at five-thirty complaining of terrible pain in her left knee. She was also having withdrawal from the morphine and was shaking so hard I could barely get her to the bathroom. Ruth came out that afternoon to sit with Margo after the home care worker left. I went to Lundar to pick up a prescription for a stronger dosage.

We thought with the stronger dose of long-acting morphine at night, we'd have a good night's rest but at three-thirty yesterday morning Margo woke with horrible pain in her legs. Her legs are becoming swollen around the knees and down into her feet. Dr. Steyn says the liver must be failing now so we don't know how much longer she's going to be with us.

I got her to the bathroom and back to bed and then gave her the allowable dose of short-acting morphine. A short while later her pain eased enough to allow her to sleep some more. I was too emotional to sleep by then so stayed up and did some emails to friends. I knew we had to do more to make her comfortable here so I phoned Dr. Steyn's house and left a message. She'd told me she does house calls if she can get time and to call if we needed her.

Margo was still in a lot of pain by the time our home worker, Pat, arrived at nine o'clock. I took the cell phone with me when Ruth and I went to fix the fence around the lake pasture. A bit of warm weather and we'll have the heifers on grass. We finished making the rounds after pushing in close to fifty posts with the loader. Ruth was right behind me stapling wires.

I was so happy to come in the house to see Margo all washed and dressed and sitting up on the couch. She'd got ahead of her pain again and Pat had done a great job with her. I can't say enough about the wonderful caring workers we have helping us. Twenty minutes later, Dr. Steyn arrived with her eyebrow ring in place and her hair combed, but looking oh, so tired. The three of us had a conference and it didn't take her long to set us at ease because she knows exactly what to do.

We've been asking home care for a commode chair but it hasn't arrived. Elzette called the hospital and asked to borrow a commode chair until the other one arrives this week. We'll also pick up a wheelchair. She told me to double up on the dosage of short-acting morphine during the day to stay ahead of the pain and she prescribed cold packs for the hot, swollen knees. She told us to ask for a hospital bed to set up in the living room. Margo can be comfortable there and we can crank the head up so she can look out at her birds and the calves running and playing in the small pasture.

Ruth stayed with Margo in the afternoon while I went to church for the first time in ages. Back home I found Steven and family here. He was mowing grass and Linda was trying to keep the kids quiet out on the deck because Margo was sleeping. I knew she'd be disappointed if she slept through their visit so I took Jayden in my arms and went in and woke her up. She was kind of confused and demanded to know, "What's going on here?" She soon came around and was happy to see everybody.

We've had a very peaceful night and slept through until five o'clock. I gave Margo her pills right away and told her to stay still until they kicked in. She went back to sleep and is still sleeping now at close to eight o'clock. I think things will be better now we have a plan of action. With Margo in the living room, I'll sleep on the couch near her. If I start getting wiped out, I'll ask for more home care to relieve me at night.

Donna arrives this morning to spend the week. I hope Margo can stay home at least that long. Our real hope is that she can stay here until she passes on. I find it so much easier here, no traveling involved and when I need to cry I can let the tears fall. Margo and I are so close and it's wonderful that we can tell each other what is in our hearts. After all these years together we can almost read each other's mind but it's still good to put our deepest thoughts into words. We have our house in order and that makes it easier to face each new day.

Speaking of new days, I went out this morning shortly after five o'clock to check an old cow that is getting close to calving and had an absolutely marvellous half hour. How about pulling on your rubber boots if you have them, or an old pair of running shoes will do, and join me as I retrace my early morning steps.

A brilliant red sun is rising above the bush to the east of our fields and already the heavy coat of white frost is melting off the freshly mowed front lawn. A huge song sparrow in his bright mating plumage is sitting on the edge of the bird feeder chirping at me as if to chastise this old man for being too slow. It looks pretty nippy out there so I pull on my winter jacket and step out the back door. Coco comes running from his house and rubs against my legs, leaving a mixture of straw and dog hair on my freshly laundered pants. I push him away and scold him, telling him that I'm doing my own laundry now so he'd better smarten up. He knows I don't mean it and gives me another enthusiastic rub.

A few steps west of the house I stop and listen to all the wonderful sounds that spring brings to our area. My hearing is none too good any more but I have no trouble hearing the drops of water running over the edge of the eaves trough as the sun warms the frost on the roof. I must remember to get up there and clean out those troughs before the next rain.

There isn't a breath of wind and I can hear the waves left over from yesterday's brisk wind still washing up on our beach one mile west of here. Out in the Black Pond, an amorous bittern is tuning up his tuba and booming out "Oompa, Oompa." There are geese nesting in every little pond this year but with most of the females now on the nest, there is only the occasional call.

A huge flock of red-winged blackbirds is sitting in the trees in the south windbreak, doing their best to drown out every other noise. When they take a short break from their singing, I hear the chatter of a large flock of prairie chickens on their dancing ground in the neighbour's hayfield. A large red-winged blackbird sits on a strand of barbed wire with his wings spread out and tail feathers fanned, singing his special mating call which sounds like "Oh Girly". How appropriate! Across the road a pair of meadow larks are sitting on a bale and singing their beautiful, warbling mating call. It seems all the birds have only two things on their minds these days—singing and making love. Lucky them!

I head to the heifer pen and dump three pails of chop in the troughs. I hear the murmur of wild ducks in the little pond north of the pen so I sneak over and peek between the boards of the windbreak fence. Several pairs of blue-winged teals and a few pairs of spoonbills are paddling serenely around the pond. At first glance it seems peaceful until I realize that every group contains more males than females. Each pair is mated up already but the poor guys lucky enough to be chosen by their lady loves are hard at work. They spend a few seconds bobbing and nodding their heads at their gals and then go skittering across the pond chasing away any intruder that gets too close. Their final reward will be worth all the stress they're going through now. In another few weeks, they can go about their carefree ways while their mate hatches the eggs and raises the family.

Now watch your step here, it's still a bit wet and mucky after that big rain a week ago. Here's our little, lame, black cow hobbling toward us on her crippled back feet. She was supposed to be butchered for beef last summer but the fall before last she jumped the fence and got in with the bulls and became pregnant. Last July 9, she produced a fine black heifer calf. She raised it all winter and is now by herself waiting for her new baby, due on June 15.

Let's head over to the east pasture to check that pregnant old cow. I can see her from here but we'll go check all the calves for signs of sickness. Walk quietly though the herd so they'll know we mean them no harm. They recognize me and one by one approach to have their backs scratched. Even the first-calf heifers view me as an old friend and stand quietly while I talk to them. I give them all that special scratch in the itchy spot in the middle of their back—the place they just can't reach with their tongue.

What's this? A tiny new calf stands up on shaky legs and gazes at me through beautiful, clear, blue eyes as if to ask, "What is this giant intruder doing here?" Our last heifer has calved by herself and she did a good job too. The little fellow is dry and fluffy and his belly is distended with warm, rich milk. The heifer stands proudly but makes no threatening move toward me even when I come up close to her new baby. She's going to be another good one like the other heifers were this year.

Let's crawl through between the wires here and walk back under the big maples in the front yard. It's amazing to see these new green leaves growing on skinny branches that a short while ago were frozen as hard as granite stone. Somehow these tiny new leaves continue to thrive, in spite of frost every night for the past week. They must have some built-in antifreeze. Do you think all this beauty is an accident of nature? I believe it's entirely the Master's plan and I'm so grateful to be a part of it.

Here we are back in the house again. Thanks for walking along with me and thanks again for all of the wonderful messages. Margo isn't up to having visitors and she doesn't want to do much talking on the phone either, but she loves the emails that come in. She truly treasures every one.

Margo tells me to send her love to all of you and she hopes to get more messages soon. I have her up and on the couch now and it looks like we have a good day coming up. God bless all of you and keep you safe. Love, Dan"

A flood of replies lifted our spirits. Here's my reply to cousin-in-law, Linda.

"Hi Linda,

Margo is worsening quickly now and I've reached the point where I'm praying for God to take her Home where she'll be safe from pain and fear. I think very soon she'll be talking with your dear son, Darryl, as they laugh together about the time she saved a little boy from drowning at Long Point Beach. I know the rest of us would've saved him but she was the first one to him even though she was terrified of water. I guess that natural mother instinct kicked in big time. Love, Dan"

And still the loving messages kept pouring in until it became impossible to answer them all personally though I tried my best.

May 18, 2005 6:55 AM

Thanks, everyone, for the support. So many of you have such strong faith and it must've rubbed off on her. Margo told me an hour ago that she believes in God and is ready to go to Heaven to meet Him and her own family. She promised to tell my mum that I love her and that brought a flood of tears from my eyes. I'm going to have to start taking salt tablets soon with all the weeping I've been doing. The thought crossed my mind that she might be conning me to make me feel better but I believe she's much too honest for that.

Margo has been battling cancer for the past two and a half years and it has seemed so cruel. If she'd died suddenly of a heart attack, she wouldn't have gone through all this pain but she also would never have known how much I love her. It has given us a chance to talk honestly and for me to open my heart to her. I know she realizes how much I care about her. I pray she will soon be free from pain as she walks in Heaven with God and her family members who have gone ahead of her. God bless you all and thanks again. Love, Dan"

On May 19, I started a journal letter not knowing it would end mid-sentence half-way through page five. Here are a few entries from that last journal letter.

"Margo is getting very low now and I've reached the point sometime over the past few days that my prayers have changed. Now I'm asking God to take her Home quickly. It's terrible to watch her deteriorate and the pain she's going through is awful, especially in the morning when she's been too long without morphine. I'm not going to let that happen again as I've requested home care for an eight hour shift during the night. She'll get morphine every four hours like she does during the day.

I've been kneeling by her bedside twice a day and praying out loud for her for several weeks now. Secretly, I pray she'll be safe with God and her loved ones in Heaven long before this letter is finished but only He knows. It's so wonderful that I can lean on Him when I'm feeling down. My faith allows me to gain control of my emotions just by asking Him to make me strong.

The last few days are a blur in my mind but I'll touch on a few things. We're so thankful Donna was able to get time off work to come home. It has been a joy having her. It's good for her too as it will bring closure to her when Margo passes on. She's tried to be strong all week but last night she lost her cool when she and her mum had a good, long cry together. Tears are wonderful healers and it helps to know that someone cares enough about you to cry for you and with you. We have wonderful kids, each of them unique and special in their own way.

Craig and Carolyn had been waiting for me to call so they could say their last goodbyes to Margo. On Tuesday morning I called over there but couldn't get any answer. I knew Margo wanted to see them and she was having a pretty good morning. I jumped in the car and drove over to their place and asked them to visit. They were here a few minutes behind me. It was so touching to see them hugging and kissing their old neighbour. They stayed and chatted for about an hour.

Margo drifted in and out of sleep and when the laughter at times woke her, she would smile dreamily and nod off again. I remember Lorne Smith sitting in our living room in the same state Margo is in now. He only lasted about a week after that but every case is different and we just have to wait and see. In the meantime I treasure every breath she takes and every moment I spend with her. I don't want her to go yet but when I watch her suffer, I wish her to be in peace.

Later, same day

Things got pretty hectic in the morning with a few visitors coming to see Margo. Georgina and Raymond came because I'd called them the day before suggesting they'd better not wait too long. Georgina broke into tears as soon as she saw Margo but it was good for all of them to cry together. It was strange to see Margo being the strong one and comforting others the same way she always does. That's the way she's always been and that's why people love her so much.

Art arrived to take Donna back to the airport. Then Margo's sister, Helga, and her son, Roger, arrived and it was getting to be too much for Margo. I asked them to sit quietly and let her sleep for awhile."

It was 3 AM May 21 before I found time to tap out a Saturday morning Margo Update.

"Hello Everybody,

Once again I find myself waking early with my first thought, as always, of my beautiful wife, who through no fault of her own is going to be leaving me much too soon.

I've been thinking about how many last experiences we've been having together. These experiences have come and passed without us even realizing it was a last experience. That is too sad, so for the past few days I've watched more carefully for these precious last experiences, like the five games of Sequence we played with Albert and Hulda last Tuesday afternoon. Margo sat beside me at the kitchen table trying to ignore her pain as we played guys against gals. It was appropriate that we played those games with Albert and Hulda because they taught us the game and built our board for us.

Margo's mind was muddled with morphine but she did okay with a little help from her care worker who sat near her looking at her cards. After four games we were tied up two to two and decided to play the rubber match. We reached a point where Albert and I had a run of four. Both teams had one sequence already and one more sequence would win the game. I drew a two-eyed jack but chose not to use it. After years of fierce competition, I didn't want any part of winning this special last game. The game went on for several rounds and I drew another two-eyed jack. Now we had two runs on the go while the gals had one run of four. I was watching Albert and trying to signal him not to make that last play. There was no need to worry because I think he must've been doing the same thing I was doing. It was such a relief when Margo finally drew an eight of spades and triumphantly called out SEQUENCE. We all praised her and she looked very proud of herself.

This email will be going into her Happiness Binder but I won't be reading this part of it to her. I don't want her to realize I threw that game, at least not yet. Maybe some day after she's gone, I'll be reading in that binder and her spirit will return and peek over my shoulder and understand why I did it.

I got thinking about first experiences for us and I thought about the first time she really kissed me. That was almost forty-seven years ago. I was only sixteen years old, still wet behind the ears. We had shared one brief kiss on our first date but that only left me yearning for more. One night she let me drive her from the dance hall in Lundar to a house party a few blocks away. Before we went into the party, she hugged me closely and gave me a kiss that just about knocked my socks off. I might not have been in love at that point but I was certainly in lust and you might say the rest is history now.

Memories like those carry me through these very difficult days as we near our parting. I think about her every waking minute. It's hard to get my work done but with help from friends and family we are in pretty good shape.

Home Care arranged for a hospital bed to be shipped from the city. We set the bed up in the living room and it's super comfy for her pain-filled body as it has four inches of extra padding on top of the mattress. She can lie there and watch her birds and cattle in front of the house.

Margo is doing much better now that we have her on regular doses of morphine during the night hours. The worker has to wake her up every four hours to give her the pills but she goes right back to sleep and wakes up relatively pain free. The workers are supposed to sit near her all night. I hear some pretty healthy snoring coming from the living room and I know Margo doesn't snore. That's okay though—I gave the gal an alarm clock in case this happened.

Margo is quite lucid at times, maybe more so than I am. We have our quiet conversations and continue to make plans for my life after she's gone. Thanks again for the emails of support that have been pouring in. She still loves to have me read them to her. Neither of us wants her to go so soon but we have no choice in the matter. We have to face it as we've been doing all along. God bless you. Dan"

The ever-increasing doses of morphine were causing Margo to spend most of her time extremely groggy and wanting nothing but sleep. There were also short periods where she was very hyper and extremely sensitive to the slightest noises around her. At times like that she would become agitated unless she had someone with her constantly.

While Donna was staying with us, Margo experienced the first of a series of very bad dreams, the kind of dreams I grew to think of as narcotic nightmares. One morning Margo and I had been talking about what it was going to be like for me after she died. She looked me right in the eye as she told me she didn't want me to live alone for the rest of my life. She wanted me to take another mate but didn't want me to start running around with a lot of other women as soon as she was gone.

That afternoon, I found Margo looking extremely worried. In fact she looked almost terrified. I asked Donna about it and she told me her mum had awakened from a bad dream. I went straight to my dear wife and took her hands in mine as I asked her what was wrong. Her face crumpled and she started to sob as she blurted, "I dreamed you were running around on me already." It took several minutes to calm her.

As Margo's condition worsened, our family had been making plans, trying to set a date when we could all be together at one time. Margo had a part in planning for the day she passed on and the day we would hold a Celebration of Life service for her. She'd made several suggestions about the way she wanted things done.

On May 21, I wrote an email to our kids with the subject line, "Let's Make Some Plans."

"Dear Kids,

Yesterday Steven helped me seed and pack the barley land. While eating lunch, we talked about plans Mum and I have in mind for her funeral and burial. Mum wants to be cremated and buried at Scotch Bay Cemetery. We'll either bury her ashes just south of my Granddad and Granny Cowdery or start a new Watson row.

Steven said he'd been talking to Kevin about what would happen if Mum passes on before the prawn season is over. Mum and I have talked it over with Kevin and told him there is no way we want him to cut his season short. He has to think of his wife and kids first, and shouldn't let his boss down by taking time off in the middle of the season. Kevin told me he would fly home for the service and then rush back out west again. That sounds like a whole lot of stress for everyone concerned.

Steven suggested we delay the actual service until we can all be together, maybe in July. That is such a wonderful plan that I wish I could take credit for it. It would give all of us time to let our heartaches heal a bit before we have to lay her to rest. Steven said he wants all the little cousins to be together that day and it would be wonderful for Mum to know this was going to happen. She's been trying to get all of you guys together in one spot at one time but it hasn't happened since Steven's wedding. Mum and I talked this over last night and she really would like to see it happen.

Mum at first wanted her ashes to be scattered from end to end on this farm of ours that she loves so much. I was able to convince her that we should bury the urn and some of the ashes at the cemetery. That way we can all go there from time to time to cry for her and also talk about the good memories. That could take some time because there are lots of good memories.

We could pick one day to take all the kids with us and have our own private celebration of her life as we go from favourite spot to favourite spot on this farm, scattering the remaining ashes. Mum said we could even pack a picnic lunch and have a big wiener roast somewhere. This would do a lot to heal our hurt and if Mum's spirit is hanging around, she'll be the happiest gal around as she watches us.

Mum has a strong heart and she might fool us and be around for a long time yet. I'm trying to be hopeful here as she is failing very fast and each day that goes by, she gets worse. Her arms and legs are getting so thin and that cursed devil cancer is making her belly all swollen. All she is eating now is a little bit of Jell-O and she can't last long on that. I'm praying for God to take her Home fast and ease her pain.

I went to Lundar just now to get her short-acting morphine pills put in a bubble pack. On the way home I spotted a big bunch of marsh marigolds growing in the swamp back in the bush. I took off my shoes and socks and went out through the rain, with nettles stinging my bare legs, and brought back the last bouquet of marigolds for my sweetie. Those beautiful flowers saved our marriage when we were almost ready to call it quits.

Mum is such a fighter. For the past few years she's been fighting the very brave fight for her life so she could live to see her grandkids grow up. That isn't going to happen the way we planned but if my beliefs are right, she will be around us long after her spirit departs her earthly body. I love you guys. Please get back to me on this and give me some input. Love, Dad"

"My dear Steven,

I've been sleeping soundly for a few hours but woke up with you on my mind. I immediately decided to get out of bed and write to you. I find it much easier to write what is in my heart than to tell you in person. For years we've been able to speak more freely to each other over the phone than we do face to face so this is much the same.

All of you kids are so special to me that there is no way I could choose between you, nor would I ever want to have to do so. It's said that if you want to know who your favourite child is, imagine yourself in a boat with your kids and the boat swamps some distance from shore. You have the chance to choose one and save him or her but the rest would have to drown. I've thought about this many times over the years and I know my answer would be to hold all four of you close as we all went down together. That would be so much better than having to live with the fact you'd made a choice.

You and Kevin are so different but ever so much the same. I can see qualities of your grandparents and of Mum and me in both of you. Kevin is a happy show-off like I am, ever ready to play the fool just to make people laugh. You are quiet and shy like Mum and you both prefer to stay well back from the limelight but close enough to share the laughter.

I was much too young to be a good parent to my first two kids but things changed when you arrived. Your beautiful big brown eyes captured my heart from the first day we knew you. The fact that you had so much sickness in the first few years just made us that much closer.

When you were a couple of months old, Mum was running herself ragged trying to look after you as well as doing more than her share of work. One night she'd been feeding you your bottle in the kitchen in our old house. She'd just put you back in your crib and gone back to the kitchen when she suddenly fainted. I heard the crash and came running in to find her lying unconscious on the floor. I picked her up and started back to our bed in the front room. Kevin had been awakened by the commotion and I met him standing at the foot of the stairs looking up at me through terrified eyes. I convinced him to go back up to bed by telling him that Mummy was sick but she was going to be okay. He was a good, little guy and did what he was told. I shudder to think about what must've been going through his mind. I was too busy with Mum to have time to go up and comfort him.

I took Mum to Doc Paulson the next day. He gave her some pills to take for awhile and told her to get lots of rest. I took over all the night-time trips to your crib that fall and they were countless, as you had that silly soother that you kept losing. As soon as it fell out of your mouth, you were hollering to get it back. Time after time I had to climb out of bed and get it for you. All those trips to your bedside and cuddling you close to make you sleep formed a very special bond between us that has lasted until this day.

I'm so glad that you and your family came out last night and am happy you were able to have a quiet talk alone with Mum. The best part of all was when you and I cried together and I felt your tears mingle with mine on our cheeks. Strong men do cry, Steven, and it's nothing to be ashamed about. The tears are God's way of bringing closure to us and healing our hearts which are so heavy right now as we watch Mum fade away. I wish we could've bawled our stupid heads off together but we couldn't do that for fear of scaring the kids who were close beside us. Don't be afraid to have a good cry when you're alone with dear Linda. You'll feel much better later. We have to take turns being strong and things will turn out okay.

When Mum is having a good day, I selfishly want her to stay with me as long as possible but when I see her suffering, I pray to God to take her quickly. I'm so thankful she's been telling me she does believe in God now and she knows she's going to walk in Heaven with Him and her own family who are there already. All day yesterday, she kept telling me she's seeing someone standing in the corner of the room watching her. When she turns her head to look at them, they go away. She still remembers when her mum was dying in the personal care home in Lundar. Gramma told us one day that Daddy (your Grandpa Nordal) kept standing in the hallway looking in at her, wanting her to go Home to Heaven with him. I tell Mum the person there is her own mum, waiting for her to walk toward the bright light. Gramma will be there to help Mum find her way to her own wonderful home in Heaven. It's so great to see the smile that lights up her face when she hears that.

Well son, I'm very tired and ready to go back to bed. I want to share this email with the other kids and I know you won't mind. Remember that I love all of you, with no favourites, and together we will get through this. Love to you all and God bless you and your families, Dad"

On May 23 I was up early writing this Monday morning Margo Update.

"Hi Folks,

Margo is still hanging in here with me on the right side of the clouds. Lately, I'm not sure that we are on the right side of the clouds with all the violence and crime and man's cruelty to man getting worse every day. I look at my beautiful grandkids and fear for their safety.

Things have been better for us since Margo finally asked Christ to take her Home. She's a little nervous about what she's facing and how she's going to get to Heaven. We've been talking a lot about Heaven, what it might be like and where it's located. I told her what I've read about near-death experiences. These amazing people sometimes come back to their loved ones quite changed from their former self but they all agree it was the most wonderful experience in their life.

I was wide awake late last night so went to check out an old cow that was due to deliver. There was a beautiful full moon hanging low over the southern horizon as it tends to do in the summer months. I decided to stay there in the pasture and talk with God for awhile. I looked up at that beautiful moon and was overcome with emotion and called out to God, asking Him where He was right at that moment.

"Where are you, God? Are you with me right here or are you on that beautiful moon up there? Maybe you are on the edge of that brilliant planet hanging low over the lakeshore to the west of here? Are you on one of those tiny little stars that are so far away that I'm not really sure whether I can see them or not?"

Then I stood in silence with my head bowed in prayer. An eerie wail sounded far away out on Long Point as a lonely coyote raised its head and howled. I lifted my head and listened intently, aware that the cry was coming from a spot directly under the moon. I barely had time to take that in when another call came from the lakeshore to the west and this time the cry came from directly under that bright planet. Then, as often happens, coyotes all across the countryside joined in with their mates until there was a full fledged chorus coming at me from all directions of the compass.

Shivers ran up and down my spine as I realized this was another sign from God and it was coming to me only a few seconds after I'd spoken to Him. I became aware of the sounds of frogs singing in every pond and the sleepy murmur of birds, which perhaps, like me, had found the night too beautiful for sleep. It suddenly occurred to me that God is everywhere. He is all around us and inside us and He has His loving arms holding us close and protecting us from all evils, imagined or real.

I've already come back to God but last night I totally committed myself to Him. I intend to follow Him wherever He leads me and I'll spend the rest of my life doing my best to help people who are less fortunate than I.

We're doing a better job of managing Margo's pain but she's getting extremely weak. She's eaten very little for the last five days and nothing at all since Saturday. We make sure she drinks lots of water to keep her hydrated and keep her kidneys working. The swelling and pain she had in her legs about a week ago is all gone and there's no more need for cold packs on her knees. She can still stand with my help to use the commode beside the bed. The home care workers taught me how to put my arms around her and lift her gently while she puts her arms over my shoulders and around my neck. She teases me that we're dancing cheek to cheek like we did when we were young kids. To me, she grows more beautiful every day and my love for her grows deeper with every minute that we spend together.

She is still the ever-caring one, always thinking more about others than she does for herself. When I was tucking her in late last night, she spotted her night shift worker sitting on the couch and tried to give her the blanket off her bed, insisting she would be alright with just the sheet over her.

Her kind and caring spirit has touched so many lives. I just got a call from Ray and Betty Larson from Eau Claire, Wisconsin. They'll be out here early on Wednesday to spend some quiet time with us before heading back to attend a graduation ceremony for a grandchild. Is that great or what? Betty is a cancer survivor so she knows something about what we're going through. Lots of love and God bless you and keep you safe. Dan"

Things were about to change quickly and that evening I sent out this email with the subject line, "Our last waltz and her last ride".

"Hi Guys and Gals,

Margo took a bad turn for the worse this morning shortly after nine o'clock and we had to take her to the hospital by ambulance. She hasn't been eating and when she started throwing up lots of green bile, I quickly decided we'd done all we can for her here. Elzette had promised us there would be a bed waiting for her when she needed it.

Ruth and Steven arrived about then as we'd planned on moving the heifers out to the lake pasture. They decided to stay and go fencing while I followed the ambulance in to town. Steven and Ruth and I shared a good cry together before I had to get ready.

When it came time to transfer her from bed to gurney, I saw the attendants trying to pick her up and I stopped them because I knew the pain in her spine would be unbearable. I told them I'd move her the same way I've been putting her on the commode for the past while. She was quite incoherent but was still responding to my voice when I told her what we were going to do. I put my arms around her and she held me close with our cheeks together. I kept on encouraging her until she was on her feet. Then I gently laid her down on the gurney and she was on her way. That will be another last for us, our last waltz together.

The ambulance pulled out of the yard with me following close behind in my car. As we turned the gentle curve in our beautiful treed lane, it struck me this was going to be her last ride as she left our beloved farm. I'm so glad she's held on long enough to see the beautiful green grass of spring that she loved so much. The trees in our lane were touching branches overhead and the new green leaves formed a cool, green tunnel. Through my tears, it looked to me like the long hallway to Heaven.

The new road north of here is a mess from the recent rain so I'd told the ambulance driver to go south to the Deerhorn Line and then take Broadway north to the Town Line. They drove slowly and carefully and I could see the brake lights come on before every pot hole. It gave me time to look across our beautiful, green, alfalfa field where nothing but tall trees grew before we moved here in 1965. Another flood of tears poured down my cheeks as I remembered the countless days Margo laboured beside me as we picked roots and rocks off this very first field of ours. We'd finally saved up enough money to clear about thirty acres of bush. She was as excited as I was as we visualized wonderful days ahead on our prosperous farm. We've been talking about that a lot these last few weeks. We agree that almost all of our dreams have come true, due to hard work on our part and the grace of God to help us. I never realized how many blessings we've had until this health crisis hit us so unexpectedly.

At the hospital the wonderful nurses soon had Margo tucked in. She was a bit dehydrated but all her vital signs are fine so that strong heart of hers is going to keep her here for awhile yet. I sat with her all day and this evening she is much improved and coherent again. Steven, Hulda and Ruth are going to share the night shift so I can get a decent sleep. Love, Dan"

Within an hour I received a copy of this email that Tracy sent to her support group.

"Hi everyone,

I talked to my Dad tonight & he told me Mom has taken a turn for the worse. This morning they called an ambulance & she has been taken into the Eriksdale hospital. She hasn't been eating much lately & has so far been surviving on Jell-O & water. She'll have constant care up until the point she is no longer with us. My fear is that she won't leave the hospital & will probably fade away soon. There have been orders already given that no heroic measures will be taken to resuscitate my mom & for them to use only intravenous to keep her fluids up & manage the pain.

I talked to my dad until I cried, talked again & cried some more. I got off the phone with him, saw my hubby standing shyly off to one side but looking at me with support in his eyes. Rick held me & rubbed my back & patted/stroked my head as I let the tears flow.

This has been very difficult but of course with all your support it has been made easier. I couldn't imagine being alone. For those of you on my email list that I work with, I'm going to keep to myself somewhat tomorrow. You all know how I've been lately, just one look at your caring faces & I lose it.

My mom has been such an inspiration all my life & to lose her slowly to such a terrible disease has been unbearable. I am at a point now where I know it will be better for her that she goes to the other side to watch over her "cheeky" husband, four kids & eight beautiful grandchildren. I don't know where all these words have come from but I just felt like sharing tonight. In looking over this email I've since decided to share it with my sister Donna, who I love very much & wish lived closer, and to my brother Kevin. You were always my protector & I was so happy when you scolded Dad when I was twelve when he made me cry during my "swathing lesson", to my brother Steven because he is my big brother & I was always so proud of him (although he didn't like it very much, I was always hanging around him & hoped we could be best friends.) We were only two years apart & you are the only sibling I really have any memory of growing up with. But most importantly, I will be sending this to my Dad because his love & support has been inspirational. He has always led by example & during these last few months has really showed what the marriage vows mean. I hope I didn't cause too many tears to flow. I'll be printing this off to show my husband so he knows how much his support means to me. To my family: I love you, to my friends on this list, you are also my family & I love you just the same. Take care & I'll talk to you soon. Love, Tracy"

Another quick Margo Update May 24, 2:47 AM

"Dear friends and family,

For those of you who haven't heard, my courageous wife is almost ready to go to her final rest. Now that she has accepted Christ, I'm ready to let her go as soon as possible. She has fought a brave fight right until the end but now she is tired and wants to go Home. I know she believes in God because she looks me straight in the eye when we talk about it. In the past if she was telling me a little white lie, she couldn't do that and I couldn't face her either if I was fibbing to her. Love to all, Dan"

Margo was safe in the hospital where she'd be looked after by professionals and it felt as though the weight of the world had been lifted from my shoulders. I'd tried my best and thought I was doing so well but had been getting close to the breaking point without even realizing it, trying to look after Margo's every need and unable to sleep when she slept. The truth is that I wasn't capable of looking after her at home as much as we both would have preferred that.

TWELVE
Her Last Hospital Stay

It didn't take Margo long to start bouncing back and none of us should've been surprised. She still had her fighting spirit and though she knew she was never going to get better, she set her sights on what seemed an impossible goal. She wanted to live long enough to have all her children and grandchildren gathered around her for a family portrait. Phyllis Penny offered to take as many photos as we wanted but the big task was to try to gather everybody at one time. Margo was determined to stay alive until she could collect all her chicks under her wings.

Two days later I wrote to our support group.

"May 25 1:42 AM

Hi everybody,

Here I am once again in front of the computer in my tiny cluttered office. I find emailing and especially the Margo Updates have a calming influence on me. I feel God's presence in the room and his words flowing through my mind. I don't even have to think about what to say. My fingers are flying across the keyboard with my thoughts sometimes getting far ahead of my actions.

Margo's heart is strong and her blood work perfect, with her blood pressure reading like a teenager's. I wish she could donate her blood pressure to me when she leaves. Perhaps she will. Right now we can't ignore the deterioration in her condition that has been rapid and alarming over the past two weeks.

It reminds me of a maneuver, a downward spiral, which I learned in flying school. You pointed the plane's nose almost straight down and banked into a steep turn. The plane would rapidly pick up speed and if allowed to continue unchecked, would eventually increase to a point where the wings would tear off and the plane would self-destruct. Before things got totally out of control, you were taught to pull the throttle back and level the wings, then pull back on the yoke gently, oh, so gently, remember those fragile wings, guys. Gradually the plane would become level again and the speed would return to normal. The maneuver was both terrifying and exhilarating and once I got the hang of it, it never frightened me again.

My dear wife is in a downward spiral and there is no slowing down or leveling her angel wings. I feel it isn't going to be much longer before she leaves us for better things ahead. I selfishly want to hold her here for a few more precious days together but at the same time I know she is ready to go to God. I wouldn't want to deny her that wonderful experience. I feel excitement in my heart when I think about the journey ahead of her. Once she reaches Heaven she'll be safe and happy as she walks on strong legs with no arthritis in her knees, with a body that has no cancer, straight to the One who created her.

It's time to say goodnight again. God bless you and keep you safe. Dan"

In a previous email, I told the story of the trip Margo and I took with Archie and Bernice up to the rodeo at Swan River as was related in Chapter 2. As we'd come to expect, every email brought a flood of messages back to us. I'd been wondering if I'd been going into too much detail in my emails but the responses, especially the ones from Linda Nickel and my cousin, Fred Ross, encouraged me to do even more.

"I read your emails today, and I laughed and I cried, as you recalled things from the past. What a wonderful way to get through this difficult time. It is those memories that help us get through. That wonderful love you share will never die, because it will live on in you, your children and grandchildren. Always on our minds and in our hearts, love Elmer and Linda"

"Hello again Dan,

Just read your Tuesday update and despite the very sad and tragic circumstances of Margo's illness, I must tell you that I thoroughly enjoyed your real life story of your trip across The Narrows to Swan River. In fact, your emails that I receive through sister Carol, although very sad and tearful, are also heart-warming and inspirational as you describe the steadfast courage and new found faith of your lovely wife and also as you so eloquently describe the wonderful world that was created for us. Laura and I are glad that you have so many caring and wonderful family and friends during this difficult time. Our love and prayers continue for you and Margo and your family. May God bless you and comfort you all. Love, Fred & Laura"

After things settled down a bit, I replied to the beautiful message Tracy had sent to her support group the night her mum had entered the hospital.

"Hi Tracy,

You mentioned that you didn't know where all your words were coming from. The answer is simple; they are coming to you straight from God's heart. His love and strength have been supporting you through these very trying days. Last year, when I begged God to forgive me for turning my back on Him for all those horrible years, He immediately took me back in His loving arms. He'd been holding me close all along but I'd been resisting Him with all my feeble strength. After that beautiful night in Victoria Hospital, I was a changed man, so changed that for awhile you were afraid I'd lost my mind. I remember trying to assure you not to be afraid and to watch my actions over the next few months. Then you'd see for yourself that this is real. Where do you think my words come from? My fingers start pecking and God's words come through me to you.

I have to do chores and then relieve Steven at the hospital. Poor guy has taken the night shift for the past two nights and must be getting tired. My wonderful children and my beautiful wife have been God's most precious gifts to me. I love you all so much, Dad"

The messages continued to pour in every time I turned on the computer.

"Hi. So sorry to hear Margo has gone to the hospital, yet I'm sure it's a relief to know she's in very capable hands. I thank you so much for contacting us so that we could say goodbye. When I gave Margo my special little Angel, I said, "This little Angel is going to keep you company and will stay with you." She smiled her clear Margo smile and said, "Thank you, Adeline, I admire you so much." I said, "I love you, Margo… you are such a dear friend." She said, "I love you too."

I kissed her on the forehead and we smiled at each other... her eyes were so clear for just a few seconds... our smiles connected and then her eyes clouded over and she squeezed my hand.

Love from all of us to all of you, Adeline"

"Hi Adeline,

Thank you and Harold for being such good friends all these years. The little angel you gave to Margo rode with her to the hospital but I'm not sure where it is now. I hope it didn't get lost in the shuffle. You know how partial we are to angels. Perhaps that last visit you had with Margo was not the last visit after all. Read the Update I send next to find out why. Love, Dan"

"Thursday morning Margo Update. She's back with us again.

Hello everybody,

I have wonderful news. I arrived at the hospital yesterday morning at seven o'clock and found a miracle had taken place. My sweetie was sleeping peacefully in her bed with Steven stretched out in the recliner by her side. He'd been there since midnight but said he felt guilty since he'd been sleeping all night.

We called to Margo and she opened her eyes and smiled at us with that wonderful smile of hers. She is getting more beautiful every day. The lines in her face have disappeared, leaving her with the soft skin of a teenager. She's always had a wonderful complexion but many years in the sun had etched the signs of passing years.

I was expecting more mumble-jumble talk but instead she was perfectly lucid as she asked me what kind of night I had. As ever, her first concern was for her loved ones, never for herself. The more I get to know her, the more I realize I've been living with my own live guardian angel most of my life.

We smooched and hugged as if we couldn't get enough of each other. She was worried about her jungle mouth but I assured her I didn't mind at all. Margo continued to be quite normal and when I helped her on the commode, she seemed strong.

She remembered that Ray and Betty Larson were due in and asked me when they were arriving. They arrived on time at ten and it was great to see them again. Our last visit with them had been at their winter home in the Florida Keys in 1990.

Margo continued to make perfect sense all morning. I kept waiting for the bubble to burst but it never did. She was sharp as a new tack and asked me about all sorts of things, like, had I paid all the bills, etc. Ray and I walked to the flower shop where he bought her a huge bouquet of fresh flowers to match the one I'd bought her the day before. I think in all the years we've been together, this was only the second time I'd done that. On Valentine's Day, about ten years ago, I bought her a dozen red roses. I can't imagine why I did such a silly thing that time. I guess my love for her must've overcome my thrifty (read cheap) nature long enough to get the job done.

All morning visitors kept stopping by and I was happy to be able to invite them in. I wasn't sure whether this was the calm before the storm but wanted to take advantage of it. Her beloved cousin, Darlene, walked in, bearing flowers from her garden and they had a wonderful visit talking about old times. I was so happy I was laughing all the time, as giddy as a kid with his first car. I walked to the kitchen to ask the cook if she had any coffee for our guests. She was busy getting lunch ready but she dropped everything and a few minutes later delivered a huge carafe of fresh coffee to us.

The kindness we've been shown overwhelms me and restores my faith in all mankind. There is some good in every man and God loves us all. There are some real saints living amongst us but we don't often take time to realize that or appreciate them. We get so wrapped up in our own little world that we don't stop to think that so many good people are much worse off than we are.

Donna Smith arrived to sit with Margo and feed her lunch. Margo had been hungry as soon as she woke and had eaten real food for breakfast. Her hands were so steady, she was even feeding herself. She was disgusted with her lunch tray and told us she's starting to hate Jell-O. Donna trotted to the kitchen and came back with chicken and mashed potatoes. Margo was wolfing it down when the Larsons and I left. I left there walking on clouds. I'm so grateful for every moment I see her like this. I have my sweetie back.

We had lunch in town and then drove out to the farm. I got Betty set up on the deck in the warm sunshine with Margo's Happiness Binder. Ray settled on the couch for a nap while I went out with a song in my heart to feed my cattle. The night before last, I phoned the funeral home in Ashern to talk about preparing for her death. Now we are making plans for our forty-third anniversary this summer. I realize that might be expecting a bit too much but you never know. As cousin Jim says, miracles do happen but first you have to believe. I do believe in miracles.

We got back to the hospital by four and found her the same, a little tired but perfectly rational. Ray and Betty soon left and I settled into the chair hoping to catch some sleep before Bill arrived to sing at six. It wasn't meant to be. Tracy and her Auntie Kay arrived a few minutes later to spend a few days.

Margo ate supper with a good appetite and was sitting up in a chair. Bill and I did our singing thing and I was so happy I sang too loud which made my throat sore. I got back to the room to find Tracy in the doorway with her eyes spilling over with tears. I didn't even get time to set my guitar down before she rushed into my arms. I held her close and assured her that God will keep her mum safe. I asked her if she could feel His strength in my arms and she said she could. If any good comes out of this mess, it will be that our kids will see how strong my faith has become and maybe they will learn to love Him as much as I do.

Sorry guys. I've rambled on way too long but it's been so long since we had any good news, I felt compelled to share every detail of yesterday's wonderful experience. Dan"

"Hey guys,

I was going to skip the Update tonight but want to tell you that Margo's condition continues to improve. Yesterday was so wonderful I felt like I was going to burst. I phoned Georgina last night to give her a verbal update, as she doesn't have email yet. I told her I'd been talking non-stop all day. She said that her sister, Eunice, tells her I have a condition known in the nursing circles as verbal diarrhea. They use this term to describe people like me who are going through the trauma of terrible situations like we are experiencing right now. That set my mind at ease as I thought I might be flipping out.

I have to mention the wonderful staff at our beautiful little E M Crowe Hospital in Eriksdale. How fortunate we are to live just fourteen miles from this haven for sick people in need of special care. I know everybody from the cleaning ladies to the kitchen staff; most of them are old friends.

Our nurses, like in any hospital, are angels of mercy working twelve hour shifts. They work all day on cement floors on aching legs and sometimes bad hips. You never see anything but kindness from them, no matter how tired they are. One worked from seven in the morning until eleven last night and the place was a mad house. Our hospital is the only one with an ER open for this north-west Interlake area. Sick people are pouring in from miles away. Yesterday was particularly hectic with non-stop emergencies, some of them life-threatening. Anyone who thinks our nurses are under-worked and over-paid had better think again.

Then there are the doctors like Dr. Burnett from Lundar who runs clinics out of Lundar, St. Laurent and Lake Manitoba First Nations. He also takes his turn doing twenty-four hour on-call shifts at the hospital.

What can I say about our own Dr. Steyn, who though born in South Africa is now definitely a Manitoba girl? Yesterday she trotted around the hospital all day looking after the ER as well as doing rounds. She was constantly surrounded by chaos but remained calm and by supper time had restored order. A few minutes later she came skipping into our room looking like a beautiful model heading down the runway. She said if Margo was doing well again today, I can take her out for a three-hour stretch. I know where we will go. It should be fun to see our friends at the dance in town. Won't people be surprised to see her again?

These two dedicated doctors are the glue that binds our little hospital together. Without them, people would be driving a hundred miles or more to the closest ER. Even then there is no guarantee that you could get in to see a doctor. We owe them a mighty big debt of gratitude. Love, Dan"

"Friday evening Margo Update May 27, 2005 11:12 PM

This was another pretty good day. I got to the hospital at seven-thirty this morning to relieve Ruth who'd been sitting with Margo since midnight. Margo's cousin, Harriet, had done the evening shift and she and Margo had really enjoyed talking about old times.

When breakfast arrived she was disgusted to find clear soup and Jell-O again. The kitchen staff had got her order wrong but they'd brought a tray for me so we shared my bowl of cereal and toast.

I was feeling a bit shaky, so as soon as Tracy arrived, I took off to visit an old friend, Gert Miller. Now in her eighties she experienced her own health crisis about a month ago when we almost lost her. She has bounced back and is spry as a cricket and sharp as a tack. We talked for a few minutes before I crawled into her comfortable bed where I slept for several hours.

Gert had lunch ready for me and I ate well before rushing back to check on my sweetie. Margo seemed weaker today but was still rational and talking sense. She was ready for lunch when I left to come home. I was hoping to do paperwork and pay bills but by the time I got home, I was wiped out again. All this prowling around at night is catching up to me. I slept for another two hours and woke up feeling like I might live again.

Tracy stayed with her mum until Bubby came to take over the evening shift. Bubby insisted on staying all night until I can get in there about five tomorrow morning. She's an angel in blue jeans, with a heart the size of Texas. She and Tracy are extremely close and they're having a hard time dealing with this illness that is taking their mum from them. Bubby considers us her adopted parents and she is our special girl.

Tracy and Ruth arrived back on the farm and after supper we moved a small group of cows. I've been too busy in town to pay attention to business and yesterday found a dead calf lying in a puddle of scours. I had a minor epidemic going on and the only way to solve it was to move the cattle onto clean grounds.

Once that job was done we walked along the edge of the windbreak looking at all the trees. I suggested we check on the little spruce tree I transplanted there in 1987 to celebrate our twenty-fifth wedding anniversary. It was planted in that west windbreak a couple of days before our anniversary. I told Margo that if it survived, it meant our marriage was strong and would survive too. It was a special tree so I watered it often that summer. The roots took hold in the shallow soil among the willows and it actually grew a bit. The next few years it shot up at a respectable pace and we joked that our marriage must be getting stronger all the time, as indeed it was.

A few years later a heavy fall of wet snow toppled one of the willows and bent the little tree. I trimmed the undergrowth back to give it more room. Any solid marriage takes some extra work to keep it in shape and growing stronger. Our little tree was no exception to that rule.

I've been much too busy over the past few years to think about that tree but tonight we found a healthy, young giant at least twenty feet tall. Our tree was standing straight and tall almost to the tip which was rubbed off and bent over from the weight of a big willow which was crowding it out.

I walked back to the shop and brought the chain saw. The three of us spent half an hour cleaning around that tree to give it lots of breathing room. It is wonderful seeing Tracy and Ruth helping each other through the pain of their impending loss. They are very alike in that they try to do more for others than they do for themselves.

I can see many wonderful traits of their grandparents expressing themselves in the actions of our four children. While our kids were little we tried our best to teach them right from wrong. I say we should get an A for our efforts. All four of them are very hard workers and are raising their own families against almost impossible odds. What are the chances that all our eight grandchildren will grow up to be normal, productive citizens? From what I've seen of them so far, those chances are pretty good.

When we'd finished, the tree was standing in its own little clearing with lots of room to spread its angel wings. The tip, although a bit bent and bruised, just like my dear Margo, is still pointing up toward Heaven. After I am alone, I will go and sit beside that tree and imagine her beautiful spirit breaking free of that tired, pain-ridden body and winging its way home to her final resting place on the other side of the valley of the shadow of death and over that last high mountain. I'm getting this keyboard wet again so will call it a night. Love, Dan"

"Saturday afternoon Margo Update May 28 5:29 PM

Hello Gang,

I just woke from a two hour nap and feel refreshed and ready to go again. I'd been getting a little burned out but have managed to get some sleep lately.

I've mentioned the words "verbal diarrhea" that is used to describe non-stop talking that people in stress experience. I thought this was an unkind term and not fitting for such a sensitive issue. When I woke from my nap, I found the answer was on my lips. What about "verbal volcanoes"?

When people are under extreme stress, tension and fear builds in their brain until they are operating under severe pressure. Some people cry and release the tension, others talk non-stop to anyone who has patience to listen, and some keep it all in until they snap or explode.

Instead of the mountain (brain) being allowed to vent off steam and lava out the hole near the top (mouth), blockage occurs that prevents the natural release of pent-up energy and the mountain explodes. The normally healthy person is then only partially whole and becomes useless to be of any comfort to the loved one who needs him or her most.

Another natural release that brings me great comfort now is the power of prayer as I lean back in my Father's strong arms. I'm so thankful for the pain-filled weeks in hospitals last year that brought me to my senses and onto my knees in front of the One who created me. Friends have asked me since if I think God is real and my answer is that I know He is real. If I feel the need to be extra strong, I ask and He calms me almost immediately and I can be strong again for dear Margo.

Please, gals, check your breasts tonight if you are lucky enough to still have them. You might find a lump and with early diagnosis you could be treated and live to be a ripe old age.

Several months ago, I promised myself I would stop getting caught up in the emotional roller coaster. How well I remember the feeling in my heart the first time Margo felt the hard painful lump on the muscle of her chest wall.

In the two and a half years since Margo was diagnosed with breast cancer which had already spread to her liver, we have run the whole range of emotions from elation, "Yay, those horrible treatments are actually working," to the deepest despair when our hopes and dreams came crashing down around our ears when the next test showed the cancer had spread even more. I don't remember the highest peak or the lowest valley but I finally came to the conclusion that I wasn't going to allow this to happen to me anymore. I would try to stay on an even keel and so avoid more heartbreak.

I found out it had been an impossible goal when I came to the hospital Wednesday morning and found Margo looking and acting like her old self. I'd spent the previous four days bawling my eyes out. It just took a sympathetic glance or a kind word to set off a fresh outburst of tears.

Now here she was back again, asking me all kinds of eager questions about the farm. I watched her carefully all morning, half convinced this was kind of the last hurrah and she would soon lapse back into the confused, shaky state she'd been enduring for the past week or so. The day went by and I found myself absolutely giddy with relief and joy. Here I was being swept up that steep slope on another wild ride and powerless to do anything about it; nor did I want to stop it.

Wednesday passed and Thursday, and I was entertaining the thought that I might take her to the dance tonight for an hour or so. Elzette told us she would have the nurse unhook the intravenous line so we could doll sweet Margo up and, "Ooh, it will be so romantic."

My bubble burst when I got to the hospital yesterday morning to find her unresponsive and just as confused as she'd been before. I'd reached the valley of despair once more.

I've had all night to think and dream about it; some of my best ideas come to me as I sleep. I know now I was wrong to want to level out the path we are walking, as hard as it seems at times. There is nothing wrong with having hope spring into your heart at any little sign of improvement, and nothing wrong with crying your eyes out when things get worse.

I was fortunate being born an eternal optimist. I can see good things coming out of almost any bad situation. My dear dad taught me a little poem when I was a child and I've never forgotten it or the meaning which he carefully explained to me.

"The clouds are very black it's true, but right behind them, shines the blue."

Even on the darkest days when the storm clouds are hanging over us, we know we just need to have patience and the clouds will pass. The sun will blaze forth in all its glory and the skies will seem even bluer than they were before the storm. With God walking beside me, holding my hand and carrying me when I become too weak to go on, I know I will see my own personal storm pass and I will smile once again through my tears.

Lots of love to you all and may God bless you and your families. Dan"

"Saturday evening Margo Update May 28, 8:30 PM

Hi guys,

I spent a pleasant hour feeding cattle and checking on those sick calves. They are out on green grass and are all looking much better. For some of them, all they needed was to get away from those evil bacteria that were polluting their former pasture. Others needed a little TLC like I gave them yesterday. A couple got scour tablets and injections of antibiotics, but one new calf was so sick I had to hold him down and give him two liters of electrolytes by stomach tube. This afternoon he is dancing around and will be joining his older mates on green grass tomorrow.

Even during years when I was pulling away from God, I saw His might and power all around me in the wonderful nature that I love. The sight of a tiny, wet, newborn calf struggling to fill his lungs with clean cold air and that first mighty gasp as he experiences freedom from his mother's womb. I love to crouch beside him and marvel in his beauty, and wonder how he survived for nine months in the womb with only that slender umbilical cord as a life line. I knew deep in my heart this had to be much more than a simple act of nature.

I just phoned Gert and she says Margo is confused and missing me badly. Ruth is sitting with her for the evening hours and Georgina is coming at midnight. How blessed we are to have wonderful friends and family. I'll always be grateful to you. Love to all, Dan"

Late Saturday night I received an email from one of Tracy's co-workers and she confided that she and some of her fellow workers were planning something special for Margo.

"Hello Mr. Watson,

I just thought I'd type a line to see how things were going. I've just read your updates. It makes me happy to read the emails on her condition & read that she is doing better. It sounds like you have great support from family, friends, hospital staff & the care workers who come into your home. I want to let you and Tracy know something. On Thursday, 05/26/05, they put the Bears on Broadway from Main Street to Memorial Blvd. (they are for cancer) A few of us were checking out the newspaper & there was a booklet in there, containing all the artists who painted the bears & their meanings. We noticed a bear in there called the snow angel. It represents people who had cancer or passed on from it, to honor these loved ones. For a donation, we can have Margo's name hand painted on this bear. As a memorial for her, I thought it was a wonderful idea, I hope you're ok with this.

You're in my thoughts & prayers. Melissa"

"Hi Melissa,

That sounds like a great idea. Angels have had a special part in our lives. I'm all for the idea, especially if we could get Margo's name on the snow angel bear.

What a marvelous idea! I wonder who thinks up things like that. Probably special people like you. Love and God bless, Dan"

"Dan,

Good morning. The donation does go to cancer research. The donation is $75, but we at work will be contributing to it. We do not want you or Tracy to donate. We'll take care of the donation, for it is our gift to you in Margo's memory. We haven't done this yet, but we'll be doing it soon; the bears are going to be on site on Broadway until 10/2005. The snow bear is at the Polo Park mall, once October comes, it will be moved, but Cancer Care will be deciding on a special place to move it to, so it is kept safe. We decided on this instead of flowers, because it will always be there for Tracy & you & family & friends to see. Love & God bless, Melissa"

That same evening I sent this note to Donna's kids.

"Hi Dan and Becky,

We loved the card you sent and thank you so much for the invitation for me to live with you after Gramma passes on. I'm hoping to visit you in Edmonton, maybe this winter, but I want to continue living here in Manitoba where most of my friends are.

No, Gramma and I aren't depressed that she's going to die. You know she believes in God and she'll go to Heaven after she dies. That means her soul will live on forever and she'll be waiting for me when it's my turn to go. Gramma isn't afraid to die but she's feeling bad that she has to leave you kids much too soon. She told me to tell you that she loves you very much and is so proud of you. I want you to grow up to be good people so Gramma and I will always be proud of you. I love you and am looking forward to seeing you when you come to the farm this summer. Love, Poppa"

"Monday morning Margo Update May 30, 2:48 AM

Margo continues about the same, up one day and down the next, but we've come to expect that and just enjoy her as she is. At her best, she's like her old self, full of merry wit and concern for others. At the worst times, she's very confused as the morphine causes her to hallucinate. This gets her agitated but we have someone with her at all times to calm her fears.

The staff at our hospital is wonderful. This morning I watched a nurse and an aid wash Margo and change her bedding without even taking her out of bed. The cook, who is an old friend of ours from way back, comes to Margo's room several times a day to see what she can cook to tempt her appetite.

When I went to relieve Bubby on Saturday morning, I found Margo agitated and confused. A little TLC from me soon had her fast asleep. Bubby said goodbye and left after doing a full night shift. She told me Margo was so concerned for her as she, Bubby, sat in that big recliner. Margo insisted that she cover up with a blanket. The room was very hot but to humour her, Bubby dutifully covered up and sat sweating under the blanket until Margo fell asleep. That's what I call a true friend.

I took the pillow from the foot of Margo's bed and put it under my head as I lay back in that comfortable recliner. I was intending to catch up on sleep but was distracted by Margo's scent on the pillow. We'd been using that pillow under her knees to make her more comfortable. Now it was beside me as close as she's been for all these years. She's always had her own scent—sort of fresh like newly cut hay or sweet flowers. She never has bad body odour even if she hasn't bathed for several days. If I go more than a day doing hard work, I stink like an old boar. They say opposites attract and that certainly has been the case with us.

I pressed my face on that pillow and breathed deeply as I yearned for her with my whole body and soul. I still have a deep physical desire for her. Those times when I'm able to serve her by helping her out of bed and to her feet are my favourite times of day. I hold her warm, slim, body to me as we waltz for a few seconds and I tremble with passion. I have to resist the urge to crush her close to me like I used to. I must remember her fragile ribs because that cursed cancer is in them too.

Georgina did the night shift on Saturday from midnight until six when I arrived. She and Margo are so close. She told me it's helping to give her closure from her own mother's death to be able to serve Margo this way. As a young mother with three little kids to look after, she was unable to be with her mum very much in her last few terrible weeks.

I was planning on sharing two of the recent Updates with the congregation at our church but I discovered I'd left the binder at home. Georgina suggested I wait until seven-thirty and then Raymond could ride with me out to our farm to retrieve them. She knew this would give Raymond and me a chance for a private talk.

Raymond arrived, a big strong, sun-burned man who is not afraid of anything in this world. I watched him lean over Margo's bed while tears welled up and ran down his cheeks. He might be big and tough but he has a heart of gold. Oh, God, I love these good people! Raymond and I did have a wonderful ride together as we talked about what was in our hearts.

At ten o'clock, Donna Webster arrived from Ashern to sit with Margo while I went to church. She and her husband, Doug, have been our friends since 1968. Our busy lives keep us from visiting often but we still remain close.

Ruth came with me to church for moral support. When I shared my story, I was calm as I told my friends about plans I have for after Margo is gone. I told them I want to dedicate the rest of my life to helping others less fortunate than myself. That will help fill the void that will be left by Margo's passing.

Last week when I tried to share in church, I was a blubbering idiot, barely able to choke out the words. Yesterday, I calmly told them how I felt I'd grown and matured more in the past few weeks than I have in my entire life. I told them how I felt God had made this plan for me long before I was conceived and how He had kept me safe as a child. I also told them how I feel I could write a book that would help cancer patients. I said I felt I could help bring people into God's fold and restore faith to those who might be getting a bit shaky under the stress of facing this devil cancer. People like us, who are staring death in the face as we're doing now.

I must've talked for over half an hour without pause as the words flowed off my tongue as smoothly as they are flowing from my fingertips now. When I finished giving testimony, I read the two Margo Updates. Last week everyone was crying along with me but this week their smiles were as broad as my own.

As soon as the last hymn was sung I hurried back to the hospital to check on Margo. She was having one of her very good days. I got there in time to eat lunch with her. It was touching to see how Donna gently fed her old friend.

After Donna left, we had a quiet visit before Margo drifted into a peaceful sleep. A strange thing has happened since she accepted Christ. She's become calm and unafraid and all the lines in her face have disappeared. She looks like the girl I met so many years ago. I praise God for allowing me these last few precious days together where our love for each other has deepened to an almost spiritual level.

Ruth arrived at five to sit with Margo and I headed for the door. As I passed an open office door I spotted Debbie who'd been toiling all day at her desk. Sickness and pain don't take weekends off. This dedicated staff somehow brings order and keeps things running.

I spoke with Debbie and happened to mention that I'd come back to Christ on New Year's Eve last year. Debbie leaped to her feet and gave me a warm hug. A minute later, Gail, one of our dedicated lab technicians, gave me a hug that almost cracked a rib. They are so excited that I'm back with Christ after all these years. It never ceases to amaze me how many of our friends are firm but quiet believers in the word of God.

Why is it that we can talk about all kinds of topics but we keep our love for God hidden so deep in our hearts that most of our friends have no clue about our faith? I wasn't any better. Even after my marvelous experience with God at the Victoria Hospital, I didn't talk about my new faith to many people unless I was sure they were Christians, and even then I waited for them to make the first move.

I'm exhausted after venting my emotions for the past hour. Love to you all. Dan"

Somehow my days seemed to turn into nights and even though I would go to bed at the proper time, I would wake shortly after midnight with so many thoughts and emotions running through my head, I felt compelled to get up at once and start another Update. It was rewarding to receive a constant flow of email coming back at me.

I felt a deep conviction that I could write a book that would be a comfort to anyone experiencing the terror of cancer, or sickness of any kind, as they made their shaky way through the trials of modern medicine. It would let confused and frightened people know they are never alone. There is so much comfort to be derived from trusting God and putting faith in friends. It would let them know that it's ok to question a doctor's decision and to not be afraid to ask for answers to questions that are bothering them.

It would ultimately bring to them the real truth that if you believe in God and eternal life, there is no need to fear death. I had no way of knowing how things would unfold but it was plain for all to see that we'd reached peace and acceptance of our situation even though that might've been hard for some people to understand.

"Sent: Monday, May 30, 2005 11:28AM

Subject: From Anita (Tracy's Faneuil friend)

Dear Dan,

I wanted to remind you there are many people thinking about you & Margo. I thought that since you've laid your heart bare in your emails, I might do the same. I'm very glad for you & Margo that you can find some good in this ordeal that you're going through. When I got cancer, I got through the tough times by trying to find what good could come from it. The main things that I found were that my family ties became much stronger (the people around you suddenly realize you may not be there forever). I found out how many friends I have, who were there for me (and will be in the future if I ever need them). I was also a bit shocked because some people I thought I could count on weren't there for me (but I'm glad to know this).

I've faced my mortality & don't take for granted anymore that I'll always have time to do "things" some day because "some day" might never come. I'm telling you this because I know you understand what I'm saying about finding the good, & I trust you won't be offended. When I say this to people who've never had cancer, they don't understand my meaning & sometimes get upset with me.

From what I understand from your emails, you & Margo have had a chance to express all the things that should be said between you. This is a gift. I am a very blunt person & believe in calling a spade a spade. When I got cancer, I would just tell people outright & would be surprised that they were so upset. I found out how many people care about me, I don't know why I couldn't see this before.

I'm glad for you that you're able to express your feelings through your emails; keeping things bottled up isn't the way to do things. One day when I was going through treatments & feeling sorry for myself, Christine brought me up short by saying, "Mom, cancer didn't just happen to you, it happened to all of us." I know that I'm a stronger & better person because of my experience & I know that it's done the same for you. I hope this email isn't offensive & that you understand what I'm trying to express. I just wanted to let you know that I'm thinking of both you & your family. Please don't feel obligated to reply. With love from Anita Labossiere"

"Anita, you've looked right into my heart. You are right about everything you said. Your words tonight have touched me deeply. If you'd like to forward this to Tracy, feel free to do so. I feel my faith is having a profound effect on her life just when she needs all the strength she can muster. From your new friend in pain, Dan"

"May 31, 2005 Tuesday morning Margo Update good news 3:58 AM

Slowly but surely with a few setbacks here and there, Margo's health continues to improve. I credit that to the wonderful care and support she's receiving in our hospital.

I never was an enthusiastic reader of the Bible. How I regret that I didn't have the good book at my side throughout my life. It would've been such a comfort to me as we traveled from one crisis to another while raising our family. I must confess I haven't read it as much as I should've during the last year and a half after He allowed me back into His realm.

Lately I've been reading the story of Job to Margo and we are amazed at how the words have whole new meaning to us. We love how some of the passages seem to have been written especially for us and our long life together on our beloved farm. The other morning I was reading to her and I came to Job 5-23. "You will make a covenant with the stones in the fields. They won't keep your crops from growing. Even wild animals will be at peace with you."

I put the book down and asked, "Isn't that marvellous, honey? It sounds just like us and our farm." She looked at me with her beautiful face glowing and said three words, "Yes, it does."

She didn't have to say more. I knew what she was thinking. Our farm had more than its share of rocks when we first started farming and our kids will testify to that. We worked for years picking rocks by hand before we were able to afford machinery which made light work of that tedious job. We would seed our crops in the spring and then frantically work at picking off the biggest of the rocks before the new grain came up. It didn't really matter because the grain came up between the stones and we almost always got a fine crop. Our land is rocky but it's also fertile.

I especially like the part about how the wild animals will be at peace with you. By the time I was twelve years old, I'd taken over the care of my dad's flock of about eighty ewes with help from Margie and Ruth. I took great pride in doing even the mundane tasks, like feeding the ewes all winter to keep them in good shape for lambing. We even pumped water by hand for all those sheep.

My favourite time of year was lambing time and I was proud of my track record. The second last year we kept sheep, I never lost a lamb. The last year, I only lost one. That one was born huge and even though I gently eased him out of the birth canal, he was almost dead on arrival. I blew my breath of life into him for at least half an hour before his heart quit beating and he died. I think I cried all night.

The lambs were half grown when the flock was turned out onto green grass. Then the trouble started as the coyotes came creeping out of the surrounding bush to steal my lambs. They killed and crippled scores of them. I developed an intense hatred for those sneaky rascals and I cried for every lost lamb. The last year we were in the sheep business we lost more than twenty-five. Dad decided to sell the flock and cut his losses.

Nowadays, I think how sad God must be when He sees His precious lambs (who started as innocent babies) become possessed by the devil and lost to Him. I think I have an inkling of how He must feel.

I waged war on those cursed coyotes and for years killed them at every opportunity. I never let them go to waste; their pelts provided income for my growing family. When the price of fur dropped, I couldn't stand to kill them for no good reason. I didn't have sheep anymore and they didn't bother our calves with their protective mothers watching over them.

As I matured, I started viewing nature from a different stand-point and grew to respect pesky critters like coyotes and even the wily foxes that killed our chickens. One of the best parts of haying time is early evening when the coyotes come onto the field to hunt mice which are hiding under heavy swathes of hay. They soon realize I mean them no harm. It isn't unusual to see a pair of coyotes trotting along within a few feet of the tractor while catching mice that come scurrying out from under the baler.

Yesterday morning started off being one of God's perfect days that I feel was made just for me. As I stepped out of the house, a coyote started howling north of the yard. Another one of God's messengers talking to me, perhaps. The dogs usually go berserk with a wild animal that close to home but yesterday they stood with their heads cocked sideways listening intently. Perhaps they understand coyote talk better than I.

Driving out our driveway I was struck by the beauty of the sun peeking over the trees on the east side of our newly green barley field. Everything was dripping with dew and sparkling like diamonds. With the windows down, I breathed in the wonderful smell of flowering shrubs. As I passed an opening in the brush, I heard that coyote still singing. I backed up a bit and there she was, sitting on top of the mound of dirt left from an old brush pile. She had her head back with her muzzle pointed at the sky, yodeling as she serenaded the whole world. She likely has a litter of pups buried deep under that brush pile.

Yesterday was very good for Margo. She was moved into the palliative care room where the whole family will be more comfortable as we accompany her part way on her marvelous journey. My throat has been very sore lately and I slipped into the ER where I was checked over by Werner. He did a throat swab which was tested in no time in the lab. We in the North

Interlake owe the dedicated people of this hospital a huge debt of gratitude. If Margo was lying in some hospital or care home a hundred miles away, I wouldn't be getting to see her except for a brief visit here and there. When she wakes now in terror of being alone, there's always somebody at her side to reassure her. We who are living in rural areas must do all we can to keep our precious hospitals. My verbal volcano has been at work here. I love you all, Dan"

"Wednesday morning Margo Update June 1, 2005 3:22 AM

Good morning everybody,

The news keeps getting better and I thank God for that. Margo keeps improving and I'm dreaming again of being able to get a pass to take her home for a few hours. I'd love to be able to sit with her in our living room and share what was in my heart when I followed her ambulance out of the yard that last time.

I went in at six yesterday morning and spent a few hours with her before coming home to get the last field seeded down to barley. I had companions for Margo lined up for all day but from now on, I'll have to use a list to keep it all straight. We had three different people show up at the hospital at midnight the night before. Being the good people they are, none were upset and none wanted to hear my apologies.

Ruth and I worked like fiends all day and got the field done. She cultivated while I picked the larger rocks. We then came home and got the discs hooked up. After lunch, Ruth disked the field while I caught a two hour nap. Next, I seeded the field while she picked a huge bucket full of smaller rocks. I got the big tractor and seeder stuck on the first round and Ruth had to pull me out with the loader tractor. She dislikes towing as much as Margo but she did it and I was soon busy getting the last of this year's seed in the freshly tilled ground.

We got home for supper about six. I intended to let the ground dry for an hour and then go back out with the rock roller to pack the seeded land. The seed germinates much faster if we do that and this late in the year, time is money.

I was checking my emails, while Ruth made grilled cheese sandwiches, when the phone rang. It was my sweetie calling and she asked when I was coming back to see her. When I told her not until tomorrow morning she sounded very disappointed so I immediately told her I'd be right there. Ruth was standing at the stove with her face covered in itchy peat moss dust. I asked her if she could do the packing and she immediately said, "Shore." I've heard that same answer so many times over the past years I would've been surprised to get any other answer. Did I mention she'd also used the loader to feed the cows hay while I was seeding barley? The evening chores included slugging around about twenty large pails of heavy grain and she'd finished it all. Thanks, my dear Ruth.

Steven called to see if there was anything he could do so I asked him if he would come out and hook up the packer and finish the job. He said he'd be right out. My angels show up just when I need them most.

I'm so glad I went back to the hospital. It was a special evening as Margo and I talked for hours. She's so excited about her newly found faith and the exciting journey that lies ahead of her. I wish you could all see the amazing change. Two weeks ago she was fearful and bitter but now she's excited and buoyant as she speaks to God and asks Him for special guidance. You should hear her recite the Lord's Prayer word for word from memory. Where do you suppose that came from? She must've been listening carefully at the funerals we've attended together over the years. She's still regretful at having to leave her wonderful family and friends much too soon but then, who wouldn't be?

Georgina arrived before midnight to relieve me while I came home for a good sleep. I'm going to start a special letter to my kids now. I know they won't mind if I CC you all. Love, Dan"

"June 1, 2005 Special letter to my kids 5:50 AM

Dear Kids,

This letter is meant to be just for you kids, my special people. Please join me on another of my special walk-abouts and we'll talk as we go. You won't need boots. We aren't going far and won't even be leaving this house.

It's a beautiful moonlit night as I sit at my computer pouring out my innermost feelings. Suddenly, our two big dogs start barking loud enough to raise the dead. Worse yet, they aren't using their short, choppy coon-bark or the long drawn-out howl they use to answer their coyote cousins. This sounds like the people bark they use to warn me there are two-legged predators around. For the past two years we've been losing gas out of our bulk storage tanks, first at a steady dribble and then two hundred gallons of expensive car gas at one shot. We quit having car gas delivered to the farm and then noticed the purple farm gas was disappearing at an alarming rate. I've locked the tank and now suspect someone is out there trying to figure out how to get past this latest set-back.

Let's leave the lights off and sneak through to the kitchen to see if we can see anyone out there in the moonlight. Join hands and follow me but don't make any noise. Stick with me now because I might need some back-up out there.

I walk slowly down the hall trying not to step on that creaky board near the bathroom door. The full moon has the kitchen lit up so it's easier going now. I pass the table but I'm so intent on looking through the patio doors toward the gas tanks that I stumble over something on the floor. I know immediately I've upset the tray holding the four precious tomato plants I started from seed several weeks ago. Stay right where you are to avoid stepping on them while I take a peek outside.

The patio door slides without a sound and I step onto our beautiful deck that your mother loves so much. She hounded me for years until I gave in and built it. We haven't used it much; I was right when I told her there are too many mosquitoes and bugs here when it's warm enough for a deck party. But I have to admit, it is nice.

I stand quietly in the moonlight listening intently but there isn't a sound. Even the dogs have fallen silent now that the boss dog is out here with them. I guess it's a false alarm. Would someone turn on the lights, please?

Oh, what a mess! Our low-acid tomato plants that Margo loves are scattered with their dirt under the kitchen table. I scoop them back into their containers and I think at least three of them will make it. One of them looks like a goner. The stem is broken and it has fallen over sideways. I'll break it off and put it in a glass of water and maybe we'll have a little miracle and it will grow roots so I can replant it. Let's leave the mess; I'll sweep the floor in the morning. You four kids sit here at Granny's old kitchen table on chairs that are part of the set we inherited from my beautiful Aunt Anne Cowdery. She was a Palmer, married to my Uncle Bert and long before she died, she gave this set to your mother. She told us Mum was one of those special people who deserve special gifts. The set is well over a hundred years old and not in very good shape but I can't bear to part with it. I've always loved old stuff and old people.

My dear kids, we are facing a terrible time together, a time when we'll see your beautiful mother pass on and leave us for that better life. Try not to cry too much for her after she's gone. We have to remember she'll have no more pain from cancer or arthritis. I think God knows she has suffered here enough.

Before we reach that sad but wonderful day, I want you all to know how much I love you. I love each of you equally with all my heart and soul. I know I love you as much as it is possible for any man to love his children. I also want you to know that my love for you pales in comparison to the love that your mum has for you. I can assure you she loves you all at least a hundred times more than I do. Why do you think that is?

My theory is that after a woman carries you under her heart for nine months, she loves you long before you're born. Then she has to go through labour and childbirth. I believe the pain and suffering each woman experiences while in labour only makes her love her child more.

I was young and immature when you guys started arriving and I wasn't a kind parent to you. Nobody gave me a blueprint to follow so I just did my best. Unfortunately, I had a very sharp temper that exploded without a second's warning when your shenanigans got to be too much for me. You seemed like little devils at times but you were just mischievous, little kids enjoying life to the fullest.

I believed then and now that a certain amount of force is required to restore order from chaos but I realize now that no little kid should ever go through what I put you through in the name of proper upbringing. I still shudder to remember some of the spankings that I gave you. Do you remember me sending you out to the bushes to cut the switches that I'd use to administer the punishment? I knew that act was worse for you than the actual few lashes I gave you when you came back in. By the time I watched you cry all the way out to the bushes and back, my hard heart would melt and I could hardly bear to go through with the punishment. Many times I let it go at that and let you off with a stern warning.

After you were all grown up I realized what I'd done and have apologized to you many times and begged your forgiveness. You all say you've forgiven me for everything bad I ever did to you and I pray to God that's true.

I also thank God you had your brave mother to protect you from my explosive temper. I have to explain something to you. It was an awkward situation for all of us. Your bad behavior just about drove her crazy at times. She would insist I play the heavy and give you a good spanking. My love for you stopped me from doing that until that stupid temper exploded again and then I would get much too rough. Mum was small but she was mighty and she would spring to your defense. She could stop me in my tracks and bring me to my senses with three words, "That's enough, Dan."

Now that I have that off my chest, there's one more thing I have to talk about. I want you to know how important it is for families in crisis to stick together. You're all being supportive and loving to us but I'd like you to show more love toward each other as we get closer to the day we lose Mum. I know you are all spread out across the country. We can't see each other all the time but there is nothing stopping you from picking up the phone and calling each other. Sometimes all it takes is the voice of a friendly, concerned loved one to bring on the healing tears. You can cry on each other's shoulders over the phone, the same way you do with me.

When I was a young lad, I read a wonderful story about a man whose large family of children was squabbling over something petty. He feared they were going to leave home, never to return. He sent them out to the bush with instructions for each one to bring back one twig. His intent was not to punish them but to teach by example. He took a twig he'd brought back himself and lifting it up, he snapped it as easily as he might've snapped his fingers. Next he took the dozen twigs that his children brought back and laying them side by side, bound them with string to make a bundle. Then he dared them to break the bundle in half. None of them were able to break it. His idea was to prove to them that there is strength in numbers. Together we stand. Divided we fall.

Look at your right hand, each of you. Hold up one finger and imagine someone grabbing that finger and bending it back. Pretty darn painful, isn't it? Now hold up all your fingers and the thumb together. Now if someone grabs you, you can squeeze right back. I think you get my point. We can look at those four fingers and our thumb and imagine that it's you four kids and I, standing strong together after Mum passes on. It's going to be a terrible time but wonderful too, if her passing makes us stronger as a family.

I think that's about all I had on my mind, at least for tonight. Please pick up that phone and call each other or send a quick email of love. I know how much comfort the written word can bring. Mum's birds are singing outside the window and daylight is breaking. I've sat up half the night talking. I've just got time to shower and head in to town to see my little honey. I'll give her a kiss from each one of you. Catch you later, Dad"

I'm far from being the world's fastest typist and with my awkward, two-fingered style there was a lot of correcting to do. On looking at the time those last two emails went out, it's obvious I never slept at all that night. Luckily, the palliative care suite in the hospital had a separate room with a couch that folded out into a bed so it was possible to catch a few hours in there while Margo was sleeping. It wasn't always easy though, because of the many visitors. Margo was so anxious to see them that she didn't like it if we put a Do Not Disturb sign on the door. On the other hand, once the visitor was with Margo, I could catch up on some much-needed sleep.

"Thursday morning Margo Update June 2, 5:14 AM

Hi Gang,

Yesterday was absolutely marvellous. Perhaps God heard our prayers and realized how much we love her and need her here for awhile longer. I can picture Him checking his guest list and asking some faithful servant to fill in for Margo until He decides the time is right. I don't expect any major miracles but yesterday was miracle enough for me.

I left home before six o'clock in the pouring rain and had soon covered the six muddy miles to the pavement on the last eight miles leading to town. I was fighting sleep all the way and should've pulled over but wanted to be there on time to relieve Georgina who'd been there since midnight.

As I passed the old Hartfield schoolyard I noticed our special purple iris plants were in full bloom. Ever since we discovered the patch of beautiful blooms, we've enjoyed walking in and picking a huge bouquet to take home. These flowers have special meaning for us because this is the yard where my mum and Margo's mum went to school together. Perhaps they had a hand in planting the bulbs that have continued to spread over the years until now there are dozens of blooms.

It seems that every time I want to pick flowers for my sweetie, it has to be raining buckets. I guess if I'm too cheap to buy flowers in the store I have to suck it up and do what I have to do. My loafers were saturated and my legs were wet to the knees by the time I got back to the car but I had a huge bouquet that looked beautiful and smelled even better.

Margo was sleeping peacefully when I got there, so Georgina and I spent half an hour in the sitting room talking. That wise woman gave me some good pointers about what to do in various situations I might encounter over the next while.

After she left, I set the flowers in a vase where they would catch Margo's eye as soon as she woke. I settled in the recliner beside her with a blanket over me and was soon fast asleep. That happy state of affairs had lasted all of fifteen minutes when I heard Margo's soft voice say, "Oh, you're here." We spent the next hour hugging and smooching.

Margo is like her old self although her legs are weak from two weeks spent mostly in bed. Her sense of humour is as keen as ever and we shared many laughs. Her nurses are as happy as we are and our laughter was contagious.

Keep praying for more improvement, please, but don't let your high hopes take over your common sense. I have to keep reminding myself to enjoy one day at a time. Love, Dan"

Now that I'd given up my daily journal letters, I found myself free to sit at my desk all night pouring out my innermost thoughts and fears. Night after night, I crouched over the keyboard with the words flowing so smoothly, my fumbling fingers couldn't keep up. I'd be so totally focused, I might sit there without looking up for three or four hours straight. When my Update was complete, a new day was beginning and I'd hurry to town to see my sweetheart again.

My life formed a pattern of travel between our farm and town—days that passed slowly at the hospital as I caught precious cat naps while Margo slept, and then the trip back home in the evenings feeling refreshed and raring to go. As soon as chores were done, I'd check my inbox and read more precious words of love and encouragement. Then it would be lights out about eleven to sleep soundly for a few hours before the pattern started again.

"Friday morning Margo Update June 3, 4:51 AM

We had another great day yesterday and I spent the whole day in town with my sweetie. She continues to amaze me with her progress. Her hearing, which has always been nothing short of miraculous seems even keener now. She can usually tell I'm coming by the sound of my footsteps in the hall.

Steven woke from a sound sleep in the recliner and we chatted together for awhile before he went home. It's wonderful to see the looks that pass between Steven and his mum each time they say goodbye. There's some very serious love happening between those two.

After Steven left, I pulled a chair up near her bed so we could hold hands while we talked. I can't believe how much I love this brave lady and want to make the most of, and savour every precious moment we have left together. We can't get enough of each other and feast our eyes on each other as we talk. There is no bitterness or regret in these moments ever since the day she made peace with God. It seems she can't get enough of talking about Him and the place called Heaven. She's going on to the ultimate experience and wants to know as much as possible about it ahead of time. I keep reassuring her it is going to be oh, so, wonderful.

A beautiful thing happened to us in the afternoon. The storm clouds passed to the east and the blazing, hot sun came out in the brilliant blue skies. We held hands as we looked out at the huge, fluffy, white clouds which were looming up toward Heaven. It was so beautiful that we let our imaginations run wild as we dreamed out loud about Heaven being even half as beautiful. Margo's brown eyes were sparkling and the smile on her face was wonderful to behold. I suddenly exclaimed that I should've had a camera so we could capture this magical moment.

I jumped up and started phoning friends. I reached Big Al Kelner and asked him to take several shots of that marvelous cloud. He said he'd head straight outside and get the pictures.

Al's wife, Elsie, found out she had breast cancer about the same time that Margo's disease was diagnosed. Until then we were casual friends who might see each other at the occasional party. Misery loves company and these two courageous women soon became very close. At that point in time, Margo had been told that her cancer was inoperable and she'd been put on a series of estrogen blocker pills. Elsie had been told her cancer could be cured and she was given the option of partial mastectomy or complete removal of her breast. I believe she was leaning toward trying to save as much of her breast as possible until she talked to Margo. I happened to be in the house that day and heard Margo tell her that if she'd been given the chance that Elsie had been given, she would've opted to have it all removed. That would reduce the chance of missing some of the cancer cells and having the disease re-occur. Margo told Elsie that her breast wasn't that important. If the doctors did the job properly the first time, Big Al would still have the rest of her body and her soul to enjoy for many years. I believe Margo was deeply spiritual even then, but didn't know it. Can you imagine the courage it took for these two women to talk openly together about their horrendous troubles?

Elsie went through six terrible chemo treatments which made her so sick she thought she was going to die. Indeed, she must've often wished she was dead. All through the treatments, Margo was her rock. I think it took Margo's mind off her own mental anguish to be able to help her new friend.

After the treatments Elsie got an infection caused by her immune system being low and she almost died in the hospital. A few weeks later a complete mastectomy was performed, then came gradual recovery. It was a long, hard-fought battle but she slowly recovered strength as her hair grew back and the sparkle returned to her eyes.

When Margo started chemotherapy, Elsie was there for her and became her rock. They've formed a bond that will last to the grave and far beyond.

Well, gang, I'm going to have another nap before the alarm rings. Yesterday morning I fell asleep at the wheel on the way in to town and woke up just before the car entered a deep ditch. Luckily, I turned the wheel the proper way and recovered just in time. I wasn't traveling fast but most likely would've ended up in an ambulance on the way to the city. Love, Dan"

During that week Margo suffered another terrifying nightmare. Nurse Werner had come into our room for a chat during his morning break. We started talking about his horses and I asked if I could visit their farm and ride one of the horses. Margo wasn't saying much, just dozing in between listening to the conversation.

A short time later, I left Margo sleeping peacefully and went outside to sit in the sunshine and fresh air. On returning to our room, I found my sweetheart crying softly with a frightened look on her face. She looked at me in amazement and said, "Thank God! I thought you were dead." In her confusion it was hard to convince her she'd been dreaming and that I was quite alright. Once she calmed down, she told me she'd dreamed I'd been trampled flat by a bunch of horses and, "Oh, your face was all covered in blood." At that point her face crumpled and she sobbed as if her heart was breaking. It must've felt like that to her.

Later, I started thinking about how terrifying if must've been for her to wake up from that nightmare and find me gone. After that, I always tried to let the nurse know if I was going to be out of the room.

"Sunday morning Margo Update June 5, 2005 1:03 AM

This is God's day and if I hurry I can come up with something for sharing in church and still catch some more sleep before I join Margo at the hospital.

I'd never have believed when I got down on my knees and asked God for His help in the Victoria hospital that He would shower me with blessings the way He has. This past week, He has returned my love tenfold until it seems He guides each step and every breath I take.

I was praying for a small miracle and God must've heard me loud and clear. He also heard the prayers of everyone whose life Margo has touched and who has been praying for her since her ordeal began. He has given us the miracle we asked for and has spared her for us to treasure awhile longer.

For the past month Margo has been unable to read, so I print out the loving messages that come in daily via email and read them to her. A few days ago she asked me to give her the paper and her glasses as she wanted to catch up on the news. To humour her as I always try to do, I passed her the items and was amazed to see her reading the smallest fine print with ease. A week ago she was unable to feed herself so we spoon fed her the way we would a small child. Now she swings her legs out over the side of the bed and eats everything on her tray with no assistance. My sweetie is back with me and I thank God with all my heart for that."

"Here we go again, June 5, 2:43 AM

The day before yesterday I was able to bring Margo home to the farm for a three hour stay. It was wonderful to see the look on her face as she feasted her eyes on the sight of green grass and every tree covered with new leaves, all part of God's wonderful springtime miracle.

I wheeled Margo up that new wheelchair ramp that so far was only used to wheel her gurney out to the ambulance for what I was sure would be her last ride. She was tired by then so I tucked her into bed and crawled in beside her. We slept peacefully side by side the way we've done for so many years. It was getting late by the time we awoke and I had to hurry to get Cinderella back home before her coach turned into a pumpkin. There was a further delay when Cinderella insisted the coach be washed before the trip back to town. I hastened to do as I was bid and we were only half an hour late getting back to the hospital.

We both agreed it had been the most wonderful day of our lives and that is saying a whole lot as we have had some very good ones together.

Yesterday morning I was out of the house by five-thirty for what has become my favourite time of day, the early morning trip to town to take over from the faithful night-shift worker. Our support group remains strong with friends and family vying to take turns sitting with Margo. I think we should call that paying our respects in advance. How much better it is to do something nice for that loved one who is ill right now, than to wait until it is too late, only to find the best you can do is buy flowers and sit crying your eyes out at the funeral. We are totally blessed with such friends and family and I want to thank them all.

I've become very comfortable talking with God just like I would with a special friend, which indeed He is. As I pulled out of our driveway, I asked Him if He had another wonderful sign waiting for me.

A mile up the road, a bank of heavy fog covered the land. It was an eerie feeling when I entered that cloud hanging low over the earth. It became extremely dark down near the ground but up above I could see glimpses of blue sky. A couple miles farther, the fog partially lifted and there was the occasional flash of brilliant sunshine.

Far ahead I could see an animal on the road. It looked too tall to be a dog or coyote and I was hoping it would be a deer. As I got within two hundred yards, I could see it was a small deer, so skinny and weak I had to wonder how it had survived the winter. As I slowed the car to a crawl, a small dark object on the road became a sassy, fat crow that was strolling around the deer as if it was looking up into its face. When the crow noticed the car creeping too close for comfort, it flew away but the deer stood mesmerized, gazing east into a field of grass. It was so intent on what it was doing that I was able to approach within a few yards. It looked to be in terrible shape with shaggy hair standing coarse and rough all over its body and ribs showing stark through its thin hide. Its belly was gaunt as if it was on the verge of starvation.

Still taking not the slightest notice of me, the deer turned to walk away. My breath caught in my throat when I realized this deer, just a mere baby herself, had recently given birth to a new baby of her own. The snow white hair on her rump was stained with dried blood and membranes and dark streaks of blood covered her back legs. The udder, swollen with life-giving milk, was distended against those skinny back legs, forcing the poor creature to walk with an unnatural gait.

The sound of tires on gravel suddenly caused her to take notice. Three bounds carried her across the ditch into the field. The bounds slowed to a stiff prancing gait and suddenly this forlorn creature was transformed into a thing of real beauty. She pranced across the field with each front leg arched and her white tail waving. Reaching the safety of the bushes, she swung around to face me directly and I held my breath, afraid to break the spell. The ignition was quickly switched off and quiet reigned as I recalled the wonderful words in Job 5: 21-23. I'm amazed at how much comfort can be derived from these simple messages in times of stress and great need.

The little deer having concluded that I meant her no harm, trotted back across the field to a spot on the other side of the ditch. She looked again at the car and stamped a slender front hoof several times in warning to anyone who might want to do harm to her hidden baby. Her huge ears came forward and she looked intently at a spot in the grass a little off to one side. After a moment she walked to that spot and her head went down into the grass as she nuzzled her newborn. I was hoping the fawn would stand up on shaky legs so I could catch a glimpse but it wasn't meant to be. When the doe had drank her fill of the beauty and smell of her precious gift from God, she walked slowly away from the spot and disappeared into the bush. I reluctantly started the motor and continued my drive, marvelling at what had taken place. God graces me with His love. I cannot believe He has given me, a mean old sinner, the vision to read His word, the ears to hear it and the courage to share it with my friends.

I am completely drained, emotionally and physically. Sleep tight, my loved ones, Dan"

THIRTEEN
Spring Turns to Summer

Tuesday morning Margo Update Wonderful neighbours, June 7, 2005 6:55 AM

Hi gang,

Margo gets stronger every day and she's now eating more than I am. She's intent on staying as healthy as possible in order to live long enough to see all her children gathered around her. Wouldn't it be fitting if that was to happen on our forty-third wedding anniversary on July 7?

A few days ago, I noticed some swelling in Margo's lower legs and feet. The next day it was worse, so I brought it to the attention of a nurse who made a note for Dr. Steyn. Elzette popped in for a few minutes that afternoon and concluded she must cut back on the steroids to avoid what happened earlier this spring when Margo kind of flipped out on us. Yesterday, Margo had her slim legs back again.

This emphasizes how important it is to have family members watching carefully over loved ones as they lay helpless in the hospital. Our doctors are run ragged and often don't have time to take notice of little things that could escalate into something major. In this case a timely tip brought almost immediate results.

On Saturday I worked our cattle through the chutes with the help of Ruth, Steven and two good neighbours, Garry and Corrine Seidler. We kept at it steadily, giving the calves their vaccinations for various diseases and running cows through the chute in order to snap fly tags into their ears.

By noon we were waiting for Gord Defoort to arrive from Ashern with his semi trailer which sports a double deck. We had twenty-nine cow-calf pairs and one bull destined for our pasture about twelve miles from home. By mid-afternoon the truck hadn't arrived and we saw a huge, black cloud boiling up several miles east of here. Most of our weather comes in from the west so we were surprised to see a sheer wall of white water rapidly approaching across the neighbour's field. Ruth and I got soaked dashing to the house and Steven got trapped in the shop for close to an hour. When it was over, there was more than an inch of water in the gauge. I called Gord on his cell phone and told him not to come, as he'd never make it down the new grade north of here. We agreed he should come another day when the roads had dried up.

Yesterday afternoon, I headed to the hospital in time to have supper with Margo and I noticed the new grade had dried off nicely. The forecast was calling for showers today and periods of rain tomorrow so I let myself dream about being able to catch Gord in a spare moment so he could haul cattle later in the evening. Gord's wife told me he would be tied up until dark.

Our supper arrived as did Ruth, who had worked all day at the post office. I started phoning some close neighbours to try and arrange moving all the cattle in our small stock trailers. Half an hour later, my supper was left cold on the plate and five neighbours were heading out to hook their trucks onto trailers. The plan was to meet at our place in one hour. Ruth would stay with Margo until somebody replaced her for the evening shift.

I went zipping out the town line and as I drove, I started thinking about what the odds were that all my neighbours would be in the house when I called, or that none of the phone lines were busy. You might say it was mere coincidence or it could've been the fact that with all the rain lately, the farmers aren't able to get out on the land. I prefer to think it was pre-arranged by He who has been showering me with blessings beyond my wildest dreams all my life.

I'd barely gotten home when Seidlers arrived. They don't have a trailer but they'd agreed to help me bring the cattle in and sort the cows from calves once more, this time in very sloppy conditions. Before long we had Cameron Penny with his outfit, followed by Brad Blue with his old Jeep truck and even older stock trailer. A few minutes later Alex Flinta pulled in with his fifth wheel unit. At the ripe old age of almost seventy-six years he still does a lot of hauling for us. Next, Ted pulled in with his new twenty-foot gooseneck trailer and last to arrive was Craig Miller with his small stock trailer.

I can't begin to tell you how much it meant to me to see that long line-up of trucks waiting their turn to load. My heart swells with gratitude and my eyes overflow with tears just thinking about it.

Garry and Corrine walked out to the feedlot to help me round up the small herd. This went like clockwork with old Smoky, my German Shepherd, working wonders to keep the herd together and traveling in the right direction. We soon had the cows separated from their calves. The sorting pen was a mess with water everywhere, but firm footing underneath. As usual I found that haste makes waste. Before the loading was finished I managed to get kicked three times, once on my left elbow, once behind my left knee with a blow that almost drove me face forward into the muck, and once on the back of my right hand. This morning my elbow is swollen and sore.

Soon we were loaded and on our way in a cavalcade that must've had the neighbours talking all the way from here to Eriksdale. Cameron was in the lead and he soon left us in the dust or would have if there'd been any dust around. I was in front of Alex who had the biggest load of nine cows and I just stayed far enough ahead to keep out of his way.

When I pulled up at the gate, Cameron had backed into the lane already and was unloading his trailer. In no time the trucks had unloaded. Brad said he'd drop his truck off at home and ride with Alex to pick up the last eight cows we'd left behind. I slept in my truck until Alex got back about an hour later. I opened the tailgate and let the cows go. Alex left with a friendly wave, without even giving me a chance to thank him. Good neighbours! I can't say enough about them. God bless and take care. Hug your loved ones tonight. Love, Dan"

"Wednesday morning Margo Update June 8, 5:09 AM

Yesterday morning I arrived ten minutes late at the hospital and found Ruth almost ready to call out the Canadian troops to search for me. I'd stopped at the old schoolyard to pick another huge bouquet of purple irises. It had been so pleasant, slowly walking around, trying to pick that perfect stem with a couple of blooms freshly opened and another bud ready to burst forth.

Margo is getting stronger in many ways but failing in others as the disease progresses. The pain in her back and her left hip is worsening and she needs more morphine through the intravenous line during the day for break-through pain as they call it. She gets two morphine pills twice in twenty-four hours but that isn't quite enough to do the job. The medication muddles her thinking so that she forgets to ask for her next shot until the pain is almost out of control. Luckily, the medication gives almost instant relief so she's as comfortable as it is possible for her to be. We spent a pleasant morning together and received just enough visitors to help pass the time without tiring Margo.

I was hoping she'd feel up to cutting my shaggy hair and had brought in our barber's kit in case that would be possible. I don't know how many haircuts she's given me over the years or how much money she's saved us but the amount must be considerable. After lunch she sat propped up in bed with pillows behind her back and gave me another one. I was sure it was a last time haircut for us a month ago but with this determined lady, you just never know. I don't think she's ready to throw in the towel yet. For that I'm so very grateful.

Have a nice day, gang. Dan"

"Thursday morning Margo Update June 9, 5:19 AM

Hi guys and gals,

Yesterday was another great day except for the weather, as rain continued to pour down. It seems like we haven't seen the sun for weeks. Lots of farmers in this area have yet to turn a wheel in the fields and the new seed still sits in the bins. Some folks who did manage to get their crops in early find their seed rotting in the ground or their new plants covered in water.

For most of my life I've heard farmers described as a bunch of whiners by folks who don't know any better. These people don't understand what it feels like to have a new crop freeze to the ground in a late spring frost or to see a heavy crop of grain pounded into the ground by a hail storm late in August. We've seen hot summers where rain refused to fall for months. Sometimes the drought could change almost overnight to a flood where the rain pelted down for days at a time until the sparse crop rotted where it lay in the wet fields.

I suppose we do tend to whine a lot, though we'd much rather call it complaining about that darn weather. As we mature we find it doesn't do us a bit of good to complain and so we have developed a "next year" attitude. We might lose money on this year's grain or cattle crop but, by golly, just wait until next year. The sun is going to shine every day except when the rain arrives exactly when needed. Those dismal cattle prices are going to improve and those greedy meat packers are going to develop a conscience. The chain stores are going to charge a fair price. The consumers will go back to eating as much meat as they'd like.

In reality, we have young farmers up to their ears in debt, most likely crying in their houses watching the rain slashing down through the cold, dark night. Their bankers aren't going to wait much longer for that payment that was due the fall before last. We have areas in western Manitoba, stretching from the United States border up to Swan River in the northwest corner of our farming area, that received over a foot of rain last week with another foot falling in the past two or three days. Who knows where it will end? When does that young farmer give up and go out to find a real job where he actually gets paid for the work he does?

What about the young people, like my children and their mates, who've entertained the idea of taking over the family farm when the old folks decide to retire? That day is coming up fast and they find themselves looking back at years where they watched their parents struggle against weather and other factors totally out of their control. Will they decide it's probably best to stick with what's comfortable to them in order to be reasonably secure?

It wasn't raining when I got to the hospital so we decided it would be nice if Margo could come home with me for the day. After breakfast, she was given a wash and bundled into a wheelchair and we were off in our carriage once more. We headed to Lundar so I could pick up a new cartridge for the water pump in the basement.

In Lundar, I shopped while Margo conversed with friends through the partially open window. Inside the store, Dot Johnson asked me if she could send flowers to Margo at the hospital. When she learned Margo was in the car, she put her purchases back on the shelf and dashed down the street to another store where she bought a huge bouquet of fresh flowers. When Dot came out with those brilliant blooms, it put an equally radiant smile on Margo's face.

It was wonderful to see another friend, Wendy Sigfusson, looking strong and healthy after winning her own battle with cancer.

On the way home we stopped at a repair shop owned and operated by Ralph Dodge. I'd dropped off a twisted grapple fork tooth there on our way south and I found him putting the finishing touches on it in his huge hydraulic press. He'd been struggling with it all the time we were gone. When I asked him how much I owed him, he smiled and told me he'd tack a couple dollars onto the next repair job. The kindness we're being shown on all sides blows me away and I have to wonder what we ever did to deserve it.

Margo and I had a great day at home and it was made even better when Tracy and Dylan arrived to spend a few days. Ruth came out and managed to get our bedding plants in the ground just as the next downpour arrived. She stuck to it and even covered the plants with pails to protect them from that fierce wind. We cooked steaks and mushrooms for lunch and Margo ate hers with a very good appetite. I managed an hour's deep sleep beside her in our new bed before it was time to get her back to the hospital. I had her with me for eight hours with no ill effects. We hope to do this on a regular basis as soon as the weather straightens up.

Daylight is breaking. Have a happy day and we'll talk to you tomorrow. Dan"

"Another letter to my kids June 11, 10:07 AM

Hello my dear ones,

Yesterday morning Mum was feeling so well we made plans to have her come home for a sleepover. I brought her home about six and we were joined by Donna Smith and her friend, Rick. They cooked a wonderful fish supper and brought it out to us. Mum is sleeping peacefully now and we'll be doing this again as long as I can still get her in and out of a wheelchair.

For quite some time I've wanted to write to the four of you about the four tomato plants I started from seed. This spring, I planted some tomato seeds Mum had saved in an envelope several years ago. I bought six fibre pots and some potting soil and put one seed in each pot. After faithfully watering them every few days I was rewarded by the sight of a tiny, little plant struggling to emerge from the cocoon of earth that held it captive for several weeks. I'd almost given up on them when the first one appeared, so was overjoyed I'd have at least one to transplant into the garden when the soil warmed up.

Time went on and I finally ended up with four healthy plants. Another weak one died soon after it emerged and one seed never germinated at all. I carefully tended them over the next few weeks, carrying them back and forth from the living room to the kitchen to ensure they got as much sunlight as possible. One morning I reflected on how lucky it was that God had chosen to reward me with these four strong plants that would carry on the family lines, as it were, of that particular variety of tomato plant. More reflection brought to mind that these four plants might well represent you four young people and what you mean to me.

At first glance, these plants looked the same but careful scrutiny shows they are all distinctive and as unique as each one of you. Even the two empty pots have special meaning to me. One seed never germinated at all, representing the countless sperm that didn't get a chance at life here on earth. When you think of the odds, you are all very special because you were conceived in love by your mum and me and nurtured in your early years, in much the same way I've tended these plants.

The empty pot where the seed grew for a few days and then perished reminds me of the tiny baby Mum lost in the early stages of her first pregnancy. We didn't know whether it was a boy or girl but we knew we loved it already and both of us grieved deeply when the miscarriage occurred. I know Mum suffered an agony of pain in the days leading up to the loss of the baby and many days of mental stress in the following months. I was quicker to recover, perhaps because I hadn't gone through the physical pain. There was also my job to take my mind off things that might've been. The doctor tried to console Mum by telling her that most likely the baby wasn't formed correctly. This was nature's way of making sure it didn't live long enough to be born. In hindsight, I wish we'd received some grief counselling together. I was helpless to do anything except try to be there for Mum and after awhile we found that life did go on.

Thinking about the pain your mother went through giving birth to you caused me to think about the times when each of you suffered pain and how I felt helpless to do anything about it.

When I look carefully at the plants, the first thing that comes to mind is that they are all different sizes. The short, stocky one must be you, Donna. You were the first born but for some reason never grew as tall as your younger siblings. Perhaps I worked you too hard and stunted your upward growth.

Your first pain came during the three long days that your mum was in labour before the big event. You were a stubborn cuss, even then, and you had your head firmly jammed up against Mum's spine causing her untold agony. I don't know why the doctors let her labour for so long before they intervened and straightened you out. You arrived sporting a huge lump on the side of your head and a face that resembled a boiled moccasin. We loved you just the same and proudly showed you off to all who had the patience to look.

When you were about seven years old, your little brother reached a point where he'd taken enough of your bullying and he hauled off and bopped you on the nose. It bled on and off for two days, despite our attempts to stop it, using every home remedy in the book and some that weren't. The final crisis came one evening when I sat on the edge of my bed urging you to keep your head back so the blood wouldn't run out of your nostril. I guess you were swallowing all the blood that was running down your throat and finally it was too much. I'll never forget the panic I felt when I saw your eyes roll back in their sockets as your little, round head fell back over my arm. I yelled for Mum to come and we quickly got you dressed for a hurried dash to the hospital. Just before we ran out the door, I jammed a gob of Vicks Vapour Rub up your nostril and the bleeding stopped. It still wasn't bleeding when we arrived at the hospital. The nurses decided to leave things as they were and keep you in for observation overnight. The nosebleed never started again so another home remedy was born.

I remember the time you took the motorbike and had a huge crash several miles from home. Somebody brought you home covered in road rash with your wounds full of gravel. We washed the scrapes on your arms and face and put a healing salve on them. That night we took you to the doctor who prescribed antibiotics to stop the infection that had already set in.

Back to that pan of tomato plants where I notice one of them standing straight and tall just like you, Kevin. It was second to emerge but like you, it quickly outgrew its older sibling. Your birth was as easy as Donna's was difficult and I know you caused Mum the least pain while being born.

Son, I'm ashamed to say that most of the pain in your young life came from my heavy hands as I attempted to discipline and make you conform to my pre-conceived notion of how a little boy should behave. I was taught that by my dad and I suppose he learned the same from his dad. I've apologized to you many times over and you've forgiven me so I'll leave it at that. I'm so proud you've broken that cruel chain. Nobody could be kinder or gentler than you are with your two little ones.

You grew to be a wonderful hockey player, with the quick reflexes and speed necessary to excel at that game. I remember one night when we traveled to play a city team whose players were at least two years older. You guys were game and gave it your all. We were down one goal when you received a terrible body check that put you out of the game in the second period. I was the manager on your bench and I took you to the dressing room to check you over. You were crying so I knew it was pretty bad. I suggested you take your gear off and put your street clothes back on.

The pain in your back was easing up when the other players trooped in at the end of the second period with the game tied up. Your coach demanded to know who'd given you permission to take your hockey gear off. I spoke up quietly and said I had. The coach was caught up in the heat of his big game and by the end of the break he'd convinced you to play the third period, even though I was in favour of finding a doctor to take a look at your back. True to your gritty nature, you went back out and scored the winning goal with a few seconds left in the game so I guess we'll never really know who was right in that matter.

A few years later you were playing your first game with the seniors one ill-fated Sunday afternoon. A very innocent-looking pile-up saw you dragging yourself off the ice with a broken leg. We loaded you up and took you to the hospital a few blocks away. You were writhing in pain so Gail, our lab technician, came to x-ray your lower leg. When she came out of the lab I asked if the leg was broken. She replied carefully that she'd seen something there. The doctor was called and though he lived in the house beside the hospital, he never came for hours. Darkness fell and still you lay there begging me to get you something for that awful pain. I guess doctors have to sleep too and maybe he'd had a hard day but I was nearing the end of my patience. I felt helpless and guilty too, for letting you skate with the big guys that day.

I finally reached the end of my rope and went to the nurse and told her she should call the doctor again. If he wasn't there within ten minutes, I would personally escort him none too gently to the hospital where he could look after my son. He was there within a few minutes and quickly gave you a shot to knock you out. I helped him set your leg and cast it. The leg healed but as the swelling went down, the cast became loose and the doctor removed it after four weeks.

A few days later, you sneaked off to watch your team play in town. You managed to get fooling around and get the darn thing broken again. I didn't have as much pity for you when I drove you back to that same doctor for repairs. He wasn't sympathetic either and sent us to a specialist in the city. This cast stayed on for six weeks and when it came off, the doctor assured me the bone was stronger than it had been before. You'd quit school by then so you were soon put to work with a chainsaw cutting the old growth trees out of the north windbreak. You were one tough kid, sticking with the job until ten cords of wood were piled up waiting for the stove.

The next plant I see has to be Steven. This is a special plant just as you are special to us. Watching you battle for life from one day to the next did something to our hearts that causes us to think about you in that special way.

Perhaps it was the fast trip with you to a Winnipeg hospital when you were three days old and critically ill from pneumonia. Maybe it was the countless times we rushed you to hospital with high fevers during your first year of life. It might've been the time you got into some formaldehyde poison that had been carelessly stored in our old, abandoned house. You drank a fair bit and it was God's grace you survived. Dr. Baffsky filled you full of ice cold milk and then gave you something to make you throw up. I held your sturdy, little body in my arms as you shivered and shook from that icy milk. You always were an obedient little guy and that stood you in good stead that day as you struggled to take one more gulp.

The doctor wasn't satisfied he had all the poison out so we made another high speed rush to the city where you received medical treatment. While there, I received a terrible tongue lashing from an indignant doctor. He certainly let it all hang out as he gave me his opinion on anyone stupid enough to have allowed this to happen. I knew he was right and stood there quietly looking him straight in the eye until the tirade ended. I was thankful he never seized you for protective custody with the Children's Aid Society.

I often look at the scars on your body and marvel that you are still with us. I see the terrible dark scar on the ball of your thumb and the raw-looking patch on the inside of your forearm where a plastic surgeon recovered enough skin to patch your injury. They remind me of the night you were playing with friends under the benches at our local hockey arena, while Kevin played and I helped coach the team. Mum rushed you to the hospital before I even knew about the accident.

A doctor stitched the wound without sewing up the artery that was bleeding. We brought you home and within a day we noticed your whole hand seemed to be turning black. Several days later the stitches were removed and we hoped all would be well. Our hopes were dashed when the wound blew open at the school playground the next day and the artery started spurting out your life's blood in an alarming fashion. One of the teachers, Bill Mosienko, bundled you into his truck and drove you to the hospital. We received a phone call telling us to get to the hospital to accompany you in the ambulance to the Children's Hospital in the city. We got there just as they were leaving. Mum dashed in beside the driver while I managed a quick kiss for my frightened, little boy. Your brown eyes looked enormous in your pale face. You were given a blood transfusion on the way to the city where you were patched up once more. Several days later you were home with one hand bound in a huge, white bandage.

Every Friday, for months, we made that long drive to the city to get the bandage changed. That stubborn wound refused to heal even after the stitches were removed the second time. The young surgeon tried to assure us it would heal in time but we knew better. I always watched closely as the dressings were changed and it struck terror into my heart to see that wound under the doctor's bright light. Proud flesh had formed all around the open wound. It was like looking down into a small, red volcano to see that artery pulsing with every beat of your strong heart. I began urging the doctor to do surgery and fix it properly but he kept putting us off. One Friday I lost my cool. I told him I hadn't been able to get decent sleep for months because I was constantly waking in terror that sent me creeping silently into your room to make sure you were still alive. I told him how much blood you'd lost in a few minutes at the school and impressed on him the fact that if this thing burst while you were sleeping, we wouldn't know a thing about it until we found you dead in the morning.

My words must've gotten through to him because he told us to bring you back on Sunday and you'd be prepped for the operation on Monday. Mum went with you to the city and was there that evening when the artery blew out again. She said there was blood squirting everywhere and nurses and doctors were scurrying until they got the bleeding stopped. The wound was repaired properly and our nightmare was over, with only scars to tell the tale.

Do you remember me writing about tripping over that tray of tomato plants as I walked through the dark kitchen late one night? The plants went rolling under the table and a lot of dirt scattered. Three of the plants seemed to be none the worse for their experience but one had suffered a broken stem and was hanging on by a thread. I stuck it in a glass of water and waited to see what would happen. A few days later it started sending out new roots and a few days after that, I transplanted it back into the pot of soil again. It finally did thrive and sent out new branches.

That plant with the broken stem reminded me of you, Steven, when we discovered you had suffered a broken back as a small child. The strange part is that we don't know when it happened. You'd been playing hockey for years but your back never bothered you until you reached the age of thirteen and faced body checking for the first time. Every time you got smacked into the boards you'd hobble back to the bench with one hand holding your back. We never thought too much of it until we took you to the doctor. He discovered that at some point in your young life, you'd suffered a broken back which had healed with one vertebra badly out of line with the rest. He forbade us to let you take part in any sports until you'd received an operation to fuse the spine. He said it was a miracle you'd endured those body checks without becoming a paraplegic.

You were given a spinal fusion at Children's Hospital. Mum went in with you the night before while I stayed home with the other three kids helping me in the hayfield. On the afternoon of the surgery I felt compelled to be there when you came out of the anaesthetic and I arrived in time to sit with you in the recovery room. I still remember how your eyes opened and got so wide when you saw me. Your arms reached out to pull me close for a tearful hug. You were in such agony and at one point about an hour later you tried to say you didn't want that operation now. An unsympathetic nurse told you it was too late to change your mind so you might as well suck it up and be quiet. I wish I could see her for about one minute right now to give her a piece of my mind.

The operation was a success but it left you with a permanently stiff back, a huge scar on your hip where they borrowed a piece of bone and an awkward way of walking. You've never let it slow you down though, either in your job or at play with your beloved kids.

There is one tomato plant left on that tray and that must be you, Tracy. The seed must have gotten washed over to the side of the pot by overzealous watering on my part and when the little plant did emerge, it was growing up against the side of the container. It made me wonder if it was a weed. It wasn't a weed but it did get off to a shaky start, just like you did.

Your mum was a natural mother and carried you proudly like she did the other kids. About mid-term through her pregnancy she started haemorrhaging one stormy, winter night. The roads were blocked and there was no way to get to the hospital. All we knew of you was a tiny heartbeat and hearty kicks from your strong, little feet and we loved you already. I had turned away from God about eight years before that and I don't remember praying to Him. Even so, He was there with us and by morning the bleeding had stopped and you survived.

Late in July, Mum went into labour and I dropped her off at the hospital before going home to continue working in the hayfield. You were a stubborn one, like your big sister. Mum spent several days in early labour in the hospital. Dr. Baffsky tried his best but finally lost his nerve and sent her in to the city.

It became a blessing in disguise that you were born there because when a specialist checked out your bones, he found you had one very shallow hip socket. You were put in a special brace that held your legs apart for the next six months. This forced the hip ball farther into the socket until a new deeper socket was formed. If you'd been born in Eriksdale, that defect may not have been discovered until you were bigger and would've taken major surgery to repair. God was working with me all my life protecting my family.

I remember the day you were skipping around on the lawn above me as I worked on our rock wall in front of the house. You were the baby of the family and terribly spoiled. As such, you never did listen to me very well. I'd matured over the years and now didn't believe in using force in order to gain respect and attention. You kept coming back to the top of that three foot high wall until suddenly you tumbled down to my level with a large stone coming right along with you. When I picked you up, blood was pouring out of a large gash that ran up from your lip into your nose.

We took you to Dr. Baffsky for repairs. He said that because you were a beautiful, little girl who hoped to grow up to be a beautiful woman, he didn't trust himself to do the stitching. He sent us to the city to have a plastic surgeon repair the gash. The doctor there did a fine job and you'd have to look very closely now to see any trace of that scar.

Well kids, I have to get out and check my pregnant cows and then Mum and I will have waffles with ice cream and strawberries before I take her back to town. Have a great day and hug your kids and your mates. Love, Dad"

"Monday morning Update, Bad behaviour on my part June 13, 2005

Hi guys,

I was on the road to the city early yesterday morning. The liver specialist thought it was time for another CT scan to check out the blood clot in my liver. It had rained another half inch overnight and things are getting very wet. Down near the city, whole fields are standing in water and a lot of crops won't be seeded this year.

I had my scan done and drove straight back to Eriksdale Hospital where Tracy had been sitting with her mum for the day. It was wonderful to find Margo in her wheelchair playing cribbage with Tracy. She'd been sitting up all day and was looking great. She knows it does her lungs good to sit up and is determined to stay as healthy as possible. If it wasn't for the pain in her hips and lower spine, I know she'd be trotting up and down the halls all day.

About a week ago, I'd been up most of the night typing a Margo Update and daylight found me slumped in my comfortable chair watching the spell check do its magical work. That was one of the days when I couldn't keep my eyes open on the drive in to town and I was looking forward to spending most of the morning catching up on sleep.

I found Margo well rested and eager to catch up on all the news. I washed her hands and face and read her the Update before breakfast arrived. I was mentally counting the hours I was going to spend sleeping in the pull-out bed. It was hard to keep my eyes open while helping an aid give Margo a thorough body wash. It was nine o'clock before we had her tucked in for what I hoped would be a long nap. Alas, it wasn't meant to be because the morning morphine pills had kicked in. You might think someone on morphine only wants to sleep but the drug can have the opposite effect on a person. When this happens to Margo, she is super alert with all her senses working in high gear. On this morning in particular, she began finding jobs for me to do. I knew from past experience it was best to go along with the game until she wore herself out and got drowsy again.

It took close to an hour of watering her flowers, bringing her the television remote and telephone and straightening up the room. Finally she announced she was ready to sleep. I tucked her in again and fell into my bed in the next room. Within a minute I could feel myself slipping into a deep sleep until a plaintive, little voice announced she needed to pee. I'd asked her if she wanted to go before I settled her in but she'd said she was quite okay. I was getting a bit cranky as I pulled my weary bones out of bed to stagger back into her room.

We've worked out a routine for getting her onto the commode so the whole job only takes a few minutes. I put one arm around her shoulders and the other arm under her knees and swing her gently to a sitting position on the edge of the bed. Then I put both arms around her under her arms and she clings to my neck as we get her on her feet. We always have our gentle but close hug before she moon-walks her slipper-covered feet around until she is in position to sit down on the commode. That remains one of our favourite times of day and I'm sure she often calls me even when she really doesn't need to go just so she can feel me close to her. I sometimes find myself asking her if she needs to pee just so I can get another of those special hugs.

I soon had her settled again and a glance at my watch told me it was still possible to grab two hours sleep before the noon meal came. I was almost asleep once more when from very far away, I heard her ask, "Dan, are you sleeping?" I forced myself back to consciousness and replied, "No, I'm not, honey. What do you want?" My hopes for that long nap faded fast when I heard her say, "The Spectator comes out today. Could you get me one, please?" The Spectator is a small paper delivered to the Interlake area. It can be obtained free by taking one out of the large metal cabinet outside the post office. I tried to stall her by telling her I'd pick one up right after lunch. Margo wasn't about to be put off that easily. She reminded me that if I didn't get there quickly, they'd be all gone and she wouldn't get to read it. This wasn't long after she found her eyes were working again and it was still a kind of miracle for both of us. With another glance at my watch I swung my legs out of bed and headed out the door. I reminded myself it's only pay-back for all the times she went above and beyond to help me when she was bone-tired after a long day's work.

By the time I got back to the hospital, my hopes for that nap were fading fast. I proudly brought the paper to my sweetie like some great hunter bringing in a much needed meal. Margo was drowsy at last and she ordered me to put the paper on the table and that is where it sat for the rest of the day.

I stretched out and was asleep within moments. Five minutes later there was a sharp rap on the door, which had a "Do Not Disturb If Door Is Closed" sign on it. Before I could move, the door swung open and Mary Ryden stuck her head around the corner. I sat up and told her we were sleeping but by now Margo was wide awake and she called to Mary to come right in. She pulled up a chair beside Margo's bed and started to chat.

By now I was getting very hot under the collar and starting to feel a bit sorry for myself. I muttered to myself under my breath but managed to keep my cool. The women were talking very quietly and I was drifting off to sleep when another tap came on the door. This time it was Mary's husband, Edwin. He'd been for tests in the ER and was ready to head for home as soon as he collected his wife. He walked through to Margo's room and told Mary they should leave. The gals were wound up now and kept on talking with Edwin joining in every now and then and there was no more sleep for me.

I was steaming mad and finally decided this had gone on long enough. I got out of bed, walked around the corner and proceeded to tell them politely but firmly that we were trying to get some sleep and would they please leave. They got up at once and said goodbye to Margo. Mary put her hand on my shoulder and told me she was sorry for disturbing me and that she had come to offer her services on the midnight to six shift. I immediately felt guilty and blood rushed to my face in a surge of true remorse. I thanked her and offered to call her when we needed her. I went back to bed and must admit I was more concerned with getting that half-hour nap than I was about having hurt her feelings.

That afternoon, my wife brought the matter to a head in the same quiet way she has handled many similar situations in our lives. She told me I'd scared her badly when I came into the room demanding they leave immediately. I told her I was sorry for scaring her but she looked at me again and said, "I'm not talking about me. What about how you treated Mary and Edwin? You were very rude and I think you should apologize to them as soon as possible." I was still on my high horse and demanded to know what I had to apologize for. Margo gave me that level look that has cut me down to size more than once as she reminded me she'd told Mary to stop in at any opportunity. I mumbled some lame excuse and the matter was dropped, at least for the time being.

I spent the rest of the day trying to put the whole, sad mess out of my mind but the more I tried, the more it became apparent to me how rude I'd been, and to such special people. Mary is a cancer victim too. While Margo was going through chemotherapy two summers ago, we heard Mary had been diagnosed with cancer and that she wasn't doing well. Pain in her stomach had finally driven her to a specialist who discovered she had ovarian cancer with a huge tumour which would have to be surgically removed. She'd completely given up hope and wanted to get it over with and die. She cut herself off from all her friends and didn't want to see or talk to anybody. Her family convinced her to have the operation and the tumour was removed successfully. The doctors said it hadn't spread but to be on the safe side she had to undergo chemotherapy. She lost all her hair and the next time we saw her she was wearing a wig with her pretty, round face as grey as ashes.

She was recovering nicely when she developed a terrible infection from the intravenous line implanted in her arm. She ended up fighting for her life in the hospital and this stage was the worst of all for her. My face burned with shame when I recalled going into her hospital room with Bill to sing her a song to try to cheer her up. It had been so easy to walk in acting like some Good Samaritan and now I didn't have enough patience to give her ten minutes alone to chat with Margo when the two of them had been through so much together.

I tried to justify my rude behaviour by telling myself I'd been over-tired and needing sleep. That was true but whose fault was it that I'd stayed up all night typing?

I phoned Mary and apologized for my rude behaviour. That gracious lady told me there was nothing to apologize for. She even tried to apologize to me again but I wasn't having any more of that and let her know just how sorry I was. A few days later, she came in at midnight and sat with Margo until I arrived at dawn. I had a good chance to talk to her again before she left. She is one of the lucky ones whose cancer hadn't spread and I pray she will stay safe and cancer-free for many long years.

The sun is almost up and it's time for my early morning ride into town. I wonder what special things await me up the road this morning. Have a great day. Love, Dan"

"Tuesday morning Margo Update, Another flood? June 14, 3:47 AM

Hi gang,

I kept waking up in the night to the sound of rain on our tin roof. It looks like about an inch in the gauge and it's been pouring for the past half hour with no signs of letting up. Our garden, which was germinating well over the past few days, has water laying on it. It's a good thing I didn't get the spuds in the ground because that side of the garden is floating.

There is a broad strip of land in western Manitoba that received close to two feet of rain in the past few weeks and they are completely flooded out as far as any farming operations are concerned. We are a bit better off here. We took a chance and seeded our barley fields early even though the land was still a bit too wet. We had to go around the low spots but we have a beautiful crop of barley that should be sprayed for weeds this week, if only it was dry enough to get on the land. The forecast is calling for lots of sunshine once we get past the rain today. We'll keep our fingers crossed and pray for dry weather for the next month. We have a wonderful crop of hay almost ready to cut but it will take several weeks of dry weather before the land can carry the machinery.

With all this water lying around, it looks like we'll have a huge crop of mosquitoes. That leaves us with a choice of putting on bug spray with a chance of getting cancer or ignoring the bites as you go about your work and risk catching West Nile Fever. Not much of a choice, is it?

I was in the hospital by six yesterday morning to relieve Ruth who'd been sleeping in the recliner beside Margo's bed. She truly has been a good friend. The other day, I came into the room while they were talking and holding hands. Margo looked at me with her eyes suspiciously moist and told me, "I sure do love your sister, Dan." Ruth's eyes got a bit flooded too, just like our weather.

Dr. Steyn said I should be taking Margo home more because she's still too well to be sitting around in there all the time. We immediately made plans for evening passes whenever we want them and they'll look after sending her pills home with us. We decided to go with every second night for now and see how it goes. Bertha will arrive tomorrow and between the two of us, we may be able to bring her home every night.

We stayed for lunch at the hospital before we came home. Margo got settled in for a good nap on the couch with her portable phone in her lap. Later, we spent a quiet evening talking before I got her tucked into bed.

It's time to wake my sleeping beauty and give her the morning pills. Pancakes, with maple syrup hand delivered from Wisconsin, are on the menu this morning. Have a great day and hug your sweethearts if you have them near. If you don't have any close by, use that phone and call them. Love, Dan"

"Later:

My plans for a nice breakfast aren't likely to happen. When I woke Margo, she was very agitated and shaky. She's scared and wants to go back to the hospital. She's in terrible pain and I guess that by letting her sleep straight through since one o'clock this morning, the break-through pain has taken over. She's suffering withdrawal symptoms from the morphine and it takes the pills a long time to get ahead of the pain. I'm hoping she'll settle down now that she's had something to drink. With her morning pills safely down the hatch she might soon feel more relaxed. It frightens me too when she gets like this as I feel helpless to ease her fear and suffering. If need be, I'll take her out through the pouring rain and get her back to the hospital. I'll stay in touch and try to send another Update tonight. Dan"

"Later, same day:

Margo is back in the hospital where they can give her better care than I can. I won't be taking her home on any sleepover again because this was one of the most traumatic days of my life. An hour after the pills, she was hungry and felt well enough to eat breakfast. I brought her a bowl of dry cereal and milk and she started right in on it while I went to make her a piece of toast. A few minutes later she put the spoon down and quit eating.

Somehow, her first concern is still for others' welfare. She said she didn't want to make herself any heavier because I might hurt my back lifting her into the car. These little comments come out all the time and they'd be hilarious if they weren't so sad. More and more I realize I've been living with a real saint and didn't appreciate the fact.

I thought the worst was over but the pain was too much for her. She was so confused I was scared to leave her alone for even one minute. She wanted to phone the ambulance because she was afraid of the pain involved when I helped her into the car. I tried to tell her she'd have a terrible ride to town in that rough ambulance but wasn't able to reason with her at all. I did the only thing I could do and that was to move her as gently as possible into the car where she rode quite happily all the way to town. It had been a terrible experience trying to load her into the car when she kept telling me I was mean. She was right about that but I still think it was the sensible thing to do. I had her back at the hospital in the time it would've taken the ambulance to get to our place.

God was watching over us and He stopped the downpour long enough to get her out of the car and into the hospital. Margo was so relieved to be there; you could see her relaxing. The pills I'd given her at breakfast time were starting to work. The nurse quickly gave her morphine through the intravenous line in her hand and she soon was sleeping peacefully. I can't imagine what it was like for people with cancer in the days before they allowed you enough morphine to stop the pain. They must've screamed non-stop. Thank God modern medicine has progressed beyond that stage. Catch you all later, Dan"

"Sunday morning Margo Update June 19, 6:32 AM

This has been a very hectic and strange sort of week and the time has slipped by so fast. Margo had a couple of bad days but she is doing a little better now.

I spent all day Tuesday with Margo at the hospital and she gradually calmed down and slept a lot. My sister, Rose, arrived by car from BC that afternoon. Bertha arrived next day. It was wonderful to walk into that hospital room and see Margo's eyes light up when she saw her big sister. Those two love each other so much and you can see it in the special looks that pass between them.

We found Margo in a very agitated state and it took her a while to calm down. That morning, one of the nurses told us she thought it was time to put a catheter in Margo's bladder to avoid having to move her to the commode. I knew her hips were getting much worse and more painful each day so this made sense. With the cancer eating away at her bones there's a good chance her hips might break with the weight of her body. That would cause even more pain and we don't want that to happen. We agreed they could do the procedure that morning.

The catheter had been put in and had worked well but the pressure of the inflatable device caused her lots of stress and the feeling she needed to go to the bathroom all the time. She'd been so uncomfortable that they removed the catheter just before we got there. She's been okay without it and we'll just have to be as gentle as possible when we move her.

I'd been thinking that with two sisters here, I'd be able to spend a bit more time at home and get some work done but that hasn't happened. Margo hates to let me out of her sight for even one minute so I've been spending more time than ever over there. Even when she has someone she loves with her, she still wants me to be there too and I can understand that. I will do all possible to spend time near her side.

Yesterday morning I was at the hospital by six o'clock and slept in the chair beside Margo until she woke just before eight. She woke up feeling so well that we started making plans to take her home on a day pass. The nurses were overworked as usual with emergencies coming in and it was close to noon before Bertha and I got Margo loaded into our car. It was a hot day with very strong winds from the south and more rain in the forecast.

Margo got the idea we should order cheese burgers and fries and take them with us. Half an hour later we were on the way home with the smell of that good food whetting our appetites. As we drove, Margo kept saying we should park somewhere and eat in the car. I told her we didn't have anything to drink. She gave me that "you silly guy look" and told me there was lots of water in the ditches. To her, that made perfect sense. You don't win any arguments with people on morphine so I kept quiet.

We pulled up in front of the house and Margo refused to get out of the car. She told me to park in the shade of the trees and we'd open the windows and eat there in the car. I'd been looking forward to getting some work done with Bertha here to watch Margo but my plans were going astray in a hurry. We opened the windows and the car filled with huge mosquitoes that attacked us immediately. Bertha and I tried to convince Margo to let us take her into the house but there was no budging her. When she told me she was scared to have me move her into the wheelchair I knew what the problem was and backed off on my demands. She was remembering how I'd caused her so much pain a few days earlier when I loaded her into the car to take her back to the hospital. She didn't want more of that cruel treatment and I couldn't blame her for that. I ran into the house and gathered some drinks to go with our meal. We drove to the beach and sat there with the huge waves crashing on the shore a few feet from the car and that wonderful warm breeze blowing through the windows. No mosquitoes could battle that strong wind so we were able to eat in comfort.

We ended the day with a slow drive to show Bertha the houses on Lundar Beach. We made a detour around by Lundar before heading back to the hospital. I got Margo settled in before going home to get some work done.

It was eight o'clock when I got into the house and I wanted to fall into bed. I brought out a steak and started defrosting it before calling Margo's room to see how she was making out. She had company there but was still lonely for me. When she asked if I was coming back in, I told her I'd be there first thing this morning. She was so unhappy about that I said I'd see how I felt after supper. She phoned back while I was eating and sounded so sad; I relented and said I'd be there as soon as possible. An hour later I walked into the hospital again. My sweetie was so happy to see me, I was glad I'd made that extra effort. I stayed with her until she was sleeping.

A few days ago, we lost another friend to cancer. He was down the hall from Margo's room and I'd been seeing him almost every day. After he passed on, I made an attempt at writing poetry. I've never written poetry until last week when I felt the same compelling urge to write a poem for my dear Margo about the way I'll feel after she leaves me. Dan"

"Dear Ellen,

I saw you and your son standing in the hall yesterday morning. You had tears on your face so I feared the worst. You were talking to someone so I just slipped on by. We want you to know that Margo and I feel so sorry about your loss. I've known Bob since I was a kid.

I woke up about an hour ago and felt compelled to write a poem for Bob and I hope you like it. I feel these words are not my own but are coming through me from a higher power. That really helps to affirm my faith in

God and knowledge that we'll leave this world for a much better place where there won't be any pain and sickness. God bless you and your family in your time of grief. Dan Watson

My Poem for Bozo

I found you looking sad, old friend, when I came in to sing for you.
When I asked what was the matter, you said you had the flu.
You said you'd been sick all winter, with that old flu bug hanging on.
I said "You're in good hands here, Bob, and your pain will soon be gone."

You said, "That Gibson sounds just fine, Dan, I had one like it, you know."
I said, "Yeah, I know that, Bob, I still can't believe you let it go."
You said that stiff old finger, made you give up playing for awhile,
But you had a new guitar now, and you really liked its style.

I watched you gamely trying, to eat food that had no taste.
You told me you didn't like it, but couldn't see it go to waste.
You had to keep your strength up, with summer coming on.
"There's work to do and fish to catch, and I've got to mow that lawn."

Just three short days ago I heard, that a cure was not to be.
That cursed cancer was the cause, of your pain and misery.
I wish I could sing once more, to help you ease your pain.
But just last night I heard the news—we'll never meet again.

I hope that isn't true, Bob, and that we'll meet again someday,

In that home beyond the mountain, in that fair land far away.
You already know the answer, but I guess I'll have to wait,
To sing and play with you once more, in that land beyond the gate."

"Wednesday morning Margo Update June 22, 8:22 AM

Hi gang,

I wish I had good news for you. We can see changes for the worse in Margo's health and the downward spiral becomes steeper. Her pain level keeps getting worse and the morphine dosage gets increased. This results in more confusion and less lucid moments before the pain breaks through once more.

I've seen a decline in Margo's happy nature and she seems to be getting more depressed. At times, we find ourselves crying together again, something we hadn't been doing for the past two weeks since her health improved. She's told Bertha several times she wishes she could die and get it over with.

She's still able to read but how much sense she makes of things is only a guess. She does seem to be staying up to date on what's going on in the world and at times she surprises me with some of the comments she makes. Because I don't get time to watch television or read the paper lately, sometimes I have to do some checking to see if she was right about something. It usually turns out she was bang on. She still manages a good game of cribbage if her opponent does the pegging for both players. A couple days ago, we played two games and she won them both. She enjoys winning, just like her old dad did, but she also graciously takes losses with her usual good humour.

When nobody is visiting, she often passes time by picking up the list of people who we phone to sit with her on occasion. She uses the portable phone to call them. That's where the troubles start as she can't concentrate long enough to dial the right number. She'll sit for an hour patiently dialing one number after another. Quite often I can hear the recorded announcement that she should hang up and dial again but she keeps dialing. I find it so hard to sit there and let her do it herself. Sometimes I intervene and ask her if she wants me to call the number. She usually lets me do it but that is one more step down that long, steep slope so I don't offer very often. Every now and then she manages to reach a friend and they have a good talk. After she hangs up, the whole process starts again. At least it entertains her and helps pass the time which must drag dreadfully when she isn't sleeping.

She doesn't seem to have any trouble dialing our number. I usually go in about six o'clock each morning and have breakfast and lunch with her before heading home to get some work done. I get several calls from her every afternoon wondering when I'm going to come back to see her. I always head back in after supper to sit with her until she's safely asleep for another night. Haying time is coming up soon and I don't know how I'm going to manage to be in two places at once. We'll find a way to deal with that when the time comes.

Yesterday morning I was up by five o'clock and on my way to the Health Science Center by six. I picked up Bertha at Ruth's house and surely enjoyed her company on the ride. My liver specialist told me my liver is pretty well back to normal now. He's very happy with how my ulcer responded to the medication.

We were back at the hospital shortly after lunch. I was still there when Dr. Steyn came in and had a serious talk with us. She'll prescribe a very good anti-depressant to help lighten Margo's mood. She also asked if we'd like to have counselling. We both felt we could benefit from talking to someone, so Elzette said she would talk to Melanie, the United Church minister about seeing us.

I went home and got some work done and went back to the hospital after supper. The nurses told me Melanie will be in this morning. Unfortunately I won't be there as I have a man coming to install a home security system. Maybe once the system is in, I'll feel comfortable enough to spend the whole night in town with Margo.

I'd been planning on calling Kevin this morning to see how their prawn fishing season is going but he beat me to it. He says the season is being closed on June 28 and then they have several days work getting the gear put away until next season. When that is done, they'll be heading for Manitoba. Kristen had emergency surgery last night for a ruptured appendix. Now she has to fight the infection that follows but it sounds like she's going to recover just fine.

I pray Margo will hang on long enough to see all our kids and grandkids gathered around us for our anniversary on July 7. It's been about nine years since we had the whole family together and all of our eight grandchildren have arrived in the past ten years.

I remember an old song that went like this... "I want a girl, just like the girl, that married dear, old Dad." I know I succeeded. Margo reminds me so much of my own mother with her kind, gentle ways and caring nature. How lucky for me that I was able to meet her and win her hand in marriage. Now the wheel has come full circle. Just like Mum did, she is leaving here much too soon. Love, Dan"

The next email was a short note to my kids.

"June 24, 2005 9:04 PM

Mum seems to be failing quickly now and it's anybody's guess how much longer she'll be with us. I see her every day but still notice how much weaker she seems, from one day to the next. It had been three days since she'd been out of bed but last night she asked me to move her to the recliner. We made the trip okay and she ate her supper there, quite happy to be sitting up straight again. When it was time to get her back into bed, I had a hard time even with a nurse helping me.

This morning I went in early so I'd be there when she woke up. It gave me a shock when it looked like she wasn't breathing but she suddenly took a big breath and then went back to breathing normally again. After breakfast Dr. Steyn came in and I asked her right in front of Mum if she thought Mum would last until you guys are all home. She said the way things are going she'll make it no problem.

Later, a nurse told me that Mum is very ill. She said the morphine is causing her to stop breathing at night in a form of apnea. After Mum fell asleep again, I watched her carefully and she does stop breathing for about a half minute at a time. Then she takes a gasp and breathes regularly for awhile.

I don't want to alarm you, as she might last for many weeks yet but I thought it would be better to prepare you for the worst in case she passes on some night in her sleep. She's fought a very brave fight and we should all be so proud of her. I think she's getting too tired of fighting the pain to care anymore.

Tracy is at the hospital and I have Dylan with me. Tomorrow is another day. Love, Dad"

"Monday morning Margo Update June 27, 9:03 AM

After about a week and a half of good weather, our rains have returned. A vicious thunderstorm dumped half an inch on us yesterday morning. Last night a huge system moved in and it's been raining all night. There is over an inch and a half in the gauge this morning and it shows no signs of letting up. I can't imagine when we'll get started haying now as the fields are full of water again after drying up nicely in the past week.

Margo continues to hang in here, determinedly counting the days until all the kids and grandkids arrive on the sixth of July. At this stage, my selfish nature wants to keep her with me as long as possible. Sometimes I feel panic rising inside me until I want to scream, "Don't leave me yet." Most of the time, I'm able to relax and put my faith in God. Only He knows when the time will be right for her to leave us. She's done all she can do and we all know how courageous she's been. If she didn't have such a strong will to live and such a brave heart, she would've left us long ago.

Tracy and Dylan spent the weekend with us. The Boy is in his terrible twos and is quite a handful. He's not bad, just obsessed with anything that looks like a telephone. I had a quiet evening with him on the farm on Friday. He was good as gold and such good company for me. It does my heart good, especially when I put him down for the night and he gives me a big hug and kiss. There is something about an innocent kid drifting into a peaceful sleep that tugs at my heart strings.

Last Saturday I sprayed my barley fields for weeds. I was amazed at how the fields had dried up and how good the barley looks. The fields will be sitting in water after this recent rain.

I'll check the cows and then head back to the hospital. Rose is with Margo this morning but I don't want to stay away too long. Margo wants me with her every waking minute and I try to oblige. Love, Dan"

"Tuesday morning Margo Update, Always someone worse off than you are, June 28, 5:48 AM

Hi gang,

The rain finally stopped yesterday morning. It looks like a bog around here and I was feeling kind of sorry for myself. I pumped the water off the low end of the garden so the plants would have a fighting chance at survival. Next, I picked up a fifty pound block of salt and balanced it on my shoulder. With a thirty pound pail of chopped grain in the other hand, I walked out to where the cows were grazing in the home pasture. It occurred to me how lucky I was to be able to do things like this at my age.

The cows were grazing in lush grass up to their bellies and one new calf was nursing at his mother's side. There will be no shortage of pasture this summer. The big problem is going to be trying to make hay in these extremely wet conditions.

Over the course of the day, I learned what others were facing. Woodlands area got five inches of rain while Argyle got eight inches. Some town near Portage got ten inches.

A lady from Marquette reported they'd received six inches of rain in about an hour during the night. They woke up to hear this terrible roar and thought a tornado had struck. The noise went on and on until the man got up enough nerve to open the door. Instead of wind, they saw a solid wall of falling water, the likes of which they'd never seen before.

After hearing these stories, I feel very fortunate we're still in a position where things will dry up enough to make hay. The hay might be coarse but as my mum used to say, it will still make better feed than a snow ball. I realize I should count my blessings instead of feeling sorry for myself.

So, there you go guys and gals. Hug your loved ones close to you every day. Apologize for unkind words spoken in anger and above all, don't let the sun go down on silly arguments between you and those you love the most. Tomorrow morning might be too late to take those words back. Dan"

"Friday night Margo Update July 1, 11:19 PM

Hi everybody,

This has been a mixed-up sort of day starting badly but ending wonderful. Every time I think I have my emotions under control, something happens that totally shakes me up.

Yesterday morning Margo seemed to be worse than ever and wasn't responding to me at all. I can usually count on her for some kind of reaction whether it be happiness to see me or annoyance with some of the things we have to do to keep her comfortable. I think I'd prefer to have her giving me heck instead of having her so unresponsive. It's very sad to see her fading away little by little. Her face and arms are very thin but her beautiful legs are swelling badly. Her body is retaining fluid and when we have to move her in bed, the imprints of our hands remain on her skin for several minutes.

Tracy and Dylan came out last night to spend the long weekend with us. Margo perked up when she saw them walk into the room. She'd spent the whole day sleeping and breathing with those long, slow breaths that scare me. I caught myself timing them and there are times when she doesn't take a breath for about thirty seconds.

This morning I was at the hospital by six o'clock and she was sleeping soundly as she's been doing every night. I woke her shortly before eight and tried to get her feeling alert enough to eat breakfast. This time she was completely unresponsive and didn't have any appetite. Even when I coaxed her to eat in order to keep up her strength, she turned her face away from me and went back to sleep. I asked one of the nurses what she thought of the situation. She suggested I get our kids to talk to their mum on the phone. She said Margo will reach a point where she goes into a coma and after that it will be too late to have a talk.

I phoned home and asked Tracy to call Donna and have her call the hospital. A short time later Donna did call. It was impossible to reach Kevin because he and his family were on their way to Manitoba.

By noon, Margo seemed more alert and when I was putting salt on her meal, she suddenly reached out and kissed me on the elbow. She knew my elbow was still swollen and sore and she was showing concern for me. This little sign of affection just about did me in. I'd been crying all morning in the room around the corner. Now the tears really fell as I tried to help her with her meal without letting her know how horrible I was feeling. I managed to keep up a happy chatter while the tears poured silently down my cheeks.

This afternoon our cousin-in-law, Linda, came to our house to stay with Dylan while Tracy came in to see her mum. The hospital is no place for that lively two-year-old and we figured we'd all get more rest this way. Once Tracy got there, I came home and put The Boy down for his nap. He was more than ready to flake out for almost an hour. Later he rode with me as I cut all the grass that wasn't under water.

I was back at the hospital in time to have supper with my sweetie. We heard a woman playing a guitar and singing in a wonderful, clear voice down the hall. A nurse's aid asked the lady to sing a song for us. She came and introduced herself as Janet Wiebe from east of St. Laurent. She sang several songs and her husband, Donnelly, came in and joined us. One thing led to another and when she told us she was recovering from breast cancer, I remembered she and her husband had been to our church last summer. At that time, her hair was beginning to grow out as she recovered from chemotherapy.

By the time Janet left, Margo had come out of her shell. I'm so glad I went back to the hospital instead of leaving Tracy there by herself. The evening ended up being a wonderful, quiet time together with Margo talking like her old self.

Kevin phoned and said they were somewhere near Calgary and will be here day after tomorrow. They're driving a motor home lent to them by

Nat's parents, so they can camp whenever they get tired. Donna, Brad, and kids will leave Edmonton on the sixth of July and drive straight through in one day. If God is willing, Margo will be here to greet all her kids and grandkids in one group. If that happens we'll have a happy forty-third anniversary. There's a nice patio area outside the care home where we can set up a couple of barbecues and Margo won't have to go through being loaded into a vehicle. Deep inside, I still want the party to be held at our farm but if that isn't possible, we'll do the next best thing and have it right there in Eriksdale. Lots of love and best wishes to you all. Dan"

"Sunday evening Update July 3, 7:44 PM

Yesterday I slept for a couple of hours in the chair beside Margo's bed before breakfast arrived. She was very weak and confused but ate all the cereal I fed her. It seems strange to be her main caregiver now but such an honour, too. While feeding her I was thinking about the hundreds of times I saw her patiently feeding her newest baby, even when she had a ton of other work waiting for her. I used to tease her about how her own mouth would open when the spoon got close to the baby's mouth. She'd pretend to be vexed with me and would make a conscious effort to avoid doing that. A few spoonfuls later her mouth would be opening again.

Shortly after breakfast, Kevin called from somewhere east of Regina. I woke Margo and let her talk to her son, in case she left us before their arrival. When I took the phone again, Kevin was pretty choked up, which was not a good thing to be while driving a motor home loaded with your most precious possessions down a busy highway. He told me they were going to try to find some place to stop for a swim and let the kids blow off some steam and they'd be here in the evening.

I thought I was going to be all cool and collected through this last stage of Margo's battle but I'm turning into a mush ball again and spend most of my time around the corner crying.

At supper time, Margo ate a bit of soup and half a sandwich. I was eating my own supper when Kevin and Nat and their kids arrived. There was a happy reunion and more than a few tears when Margo woke and saw them. The kids were soon covering her with hugs and kisses and it wasn't long before Kristen was brushing Grandma's hair. She recovered well from that burst appendix and is running and skipping like nothing happened.

A short while after they left for home, Harriet Kaartinen came to sit with Margo. Harriet is not only a neighbour but a wonderful friend and also a second cousin to Margo. Harriet is like so many of our good friends in that she has lost her mate to this cursed cancer. It amazes me how she finds courage to come in and give tender care to my loved one. There's no shortage of hugs and words of encouragement for me and that really bolsters my spirits.

Before I left the hospital, I had a long talk with Audrey Miller, who is another neighbour. She was married to Sandy, my friend who died while we were in Florida. She was left alone to finish raising five sons who have all turned out great. When the last son left the house, she never sat around feeling sorry for herself. She went out and got her nurse's degree and is now one of the angels in sturdy running shoes working twelve hour shifts on that hard cement floor. The nurses here are so gentle the way they move Margo slowly and carefully when she has to use the bed pan. As sick as she is, Margo still wakes up and asks for help to go pee and her manners are so wonderful—I have to wonder how she does it.

I came home and spent a couple of happy hours talking with Kevin and his family and Tracy. It's wonderful to have them near me and we take turns being strong. It seems when one of us breaks down, the other ones instantly become strong so we can feed off their strength. Margo's trials have brought us much closer as a family.

I was asleep by eleven-thirty but my sleep was restless as I kept dreaming I was lost in a strange city and couldn't find my car. I woke up with the feeling something had happened to Margo. I phoned the hospital and a nurse told me Margo had awakened feeling kind of teary. Fifteen minutes later I was by her side. I've been with her all day except for a short break for church. Today she seems to be drifting farther away from us and has no appetite for food, no matter how I coax her. I told her it's okay if she wants to leave us now but I think she's trying to hang on until Donna and her family gets here. Her breathing is slow and jerky and at times she's only taking three or four breaths a minute. The kids have been taking turns coming in today and I know it scares them badly when she doesn't breathe for awhile. I find myself reaching out to touch her arm so she stirs and takes the next breath.

Dr. Steyn came in and was surprised to see how Margo's condition deteriorated over the weekend. She said to give Margo as much fluid as she'll drink and she gave me a hug that rearranged several vertebrae in my spine.

Tracy had to go back to the city tonight and Kevin is sitting with his mum. Before I left, I hugged her gently and told her I loved her while tears slid silently down my cheeks. She opened her eyes and saw me crying and at once her thin arms came up around my neck as she comforted me while I wept. She is fully aware of what's happening and still her first concern is for her loved ones. She'll be the ultimate caregiver until her last breath is drawn.

Harriet will do the evening shift and Steven will take over to stay all night so I can get some sleep. I'm going out to check the cattle and let myself relax, surrounded by beautiful nature. Maybe sleep will be more peaceful tonight. Love, Dan"

"Monday night Margo Update July 4, 10:48 PM

I was at the hospital before seven o'clock this morning to take over from Steven. He said she'd been completely unresponsive until she heard my voice and then was able to talk to both of us for awhile. She sleeps most of the time. At least she's comfortable and in no pain. Her arms and face are starting to swell.

When breakfast arrived I fed her most of her corn flakes and a piece of toast. I don't believe the food is doing her any good at this stage but it does both of us good to be able to talk a little while she's eating. Kevin arrived by himself and mentioned he'd been feeding his mum the day before and how strange it felt to be doing such a thing. I thank God Margo stayed strong and was here to see them when they arrived a few days ago. I believe this is so important to the kids and me to go through these last few days with Margo. I pray she'll hold on until Donna and Brad arrive.

The earliest we can hope to get the whole clan together for a family picture is next Saturday when Tracy and Rick and The Boy can be here. I don't think Margo is going to last that long, the way things are going now. If she's still here, we'll get that picture done even if we have to crowd in around her bed in that small room.

About ten o'clock I picked up Gert Miller and she sat with Margo while Kevin and I went to bring in the cattle at our pasture just outside of town. Steven met us there and the three of us chased the small herd of cows up to the corral. One lame cow received a dose of antibiotics to cure a case of foot rot. Our pastures are largely under water and we'll be doing a lot more of these treatments before the fields dry up again.

Back in town I fed Margo lunch while Kevin hung a bird feeder outside the window. He'd brought one of the feeders from home because he knew how much his mum loves watching birds. That was such a nice idea and I wish I'd thought of doing that myself. Kevin stayed with me for quite awhile and we took turns being strong while the other one cried. I'm so happy he can let go and cry now because it's going to make it easier for him later on. He finally left to spend time with his own family.

Elsie Kelner came in this evening to sit with Margo while I came home to get some things done. As soon as Margo heard Elsie's voice, she woke up and was happy to see her. Elsie was trying to stop her tears that were falling but Margo took no notice and was chatting away like her old self. I gave her hugs and kisses and left them together.

By nine o'clock, Kevin was heading in to spend the night with his mum and I finally found time to sit and relax. It felt wonderful to go out and do some normal work today after all this time spent in the hospital. It gets so that I sort of forget there is still a real world out there. Love, Dan"

"Tuesday morning Margo Update, July 5, 4:48 AM

Good morning,

I called Harriet yesterday to bring her up to date and ask her how she made out while sitting with Margo the evening before last. She said for the most part, she sat quietly while Margo slept but while she was feeding Margo her bedtime snack, Margo started to chuckle. Harriet asked her what had tickled her funny bone but Margo never answered.

A short while later, she was dozing off when her eyes suddenly opened wide and she looked at Harriet with a beautiful smile on her face as if she'd just found out a wonderful secret. She still had the smile on her face as she fell asleep again offering no explanation.

For the past few days, I've been seeing the same thing happen when she suddenly wakes up with that big smile. I'm curious and a little envious as I try to picture what she's seeing. Maybe she's getting glimpses of her parents who've come part way back to guide her home.

My real hope is that she's seeing Jesus for the first time and that He has His loving arms wide open, waiting to show her the glories that await her in Heaven. My tears flow freely as I sit here thinking about the incredible journey on which she's about to embark.

She seems to have no intention of letting the cat out of the bag about whatever or whoever she's seeing. When I question her, it's like a curtain closes in the back of her eyes and her lips stay firmly closed. I guess we'll have to continue to wonder until it becomes our turn to know the final truth.

Ever since Margo became ill, I've experienced feelings of guilt, feelings that I couldn't quite put a handle on. I think I've been a pretty good husband to her, a little rough around the edges at first, but all in all not that bad. So, why do I feel so guilty? It's only in the last four days that I've admitted openly to myself that the question lurking in the back of my mind is, "Why? Why was she the one to get cancer when it could've been me?" I've known all along that I would've exchanged places with her at a moment's notice if only that were possible, but life doesn't work like that. I know it's time to let go of the guilt and just enjoy every moment we have left together.

Over the past months, I've come to regard my computer-music room as my own personal chapel. A very special feeling sweeps over me as soon as I come through the door, especially late at night. In this room tonight my hand fell on a photo album made up of pictures of our grandkids when they were very little. Many photos showed Margo with her arms reaching out to coax a little one to her with that wonderful smile on her gentle face and her beautiful eyes filled with love. I flipped through the pages until I realized that maybe God needs her with Him to care for and comfort some of the little children who depart from this cruel world before their lives have barely begun.

When we were going through endless rounds of chemotherapy, it seemed that Margo's treatments always fell on one of the kids' days at the Cancer Clinic. It was so hard to look at children going through such terrible treatments. On Halloween, the parents and nurses arranged a special parade of those sick, little kids. They set off past us down the hall, clutching their bags to do some serious trick-or-treating in pre-arranged places in the hospital. We saw children dressed as angels with snow-white skin, dark hollows under their eyes and not a hair on their beautiful heads. Margo and I looked at each other with tears in our eyes and huge lumps in our throats. We even saw a tiny baby not even old enough to walk, with a large tumour on the side of his head and we asked ourselves, "Why does this happen to sweet little kids barely out of their mother's womb?" None of us know the answers.

It gives me such pleasure to imagine Margo waiting there in Heaven. It's easy to visualize her being one of the special angels taking children into her loving arms and making them feel safe and secure. I've never seen a child who she couldn't charm in less than a minute. Even shy ones afraid of strangers are drawn to her and their arms reach out to her as their surprised parents watch in total amazement.

I believe God brought me out of my bed just now and into my chapel in order to strengthen my faith and give me hope for Margo's future with Him. I'll cling to that thought through the days ahead and for all the years to come. I pray with all my heart that God will bless all of you tonight and keep you and your loved ones safe. I know what some of you have been through and what some of you are facing. Put your faith in Him and you will be safe, no matter what happens. Keep in mind we are all part of the Master's plan and He knows what is best for us. Love, Dan"

"Tuesday night Margo Update July 5, 2005

As strange as it may seem, this has been a rather wonderful day. By the time my last Update was on its way, it was time to head for town to take over from Kevin.

The sun was peeping over the trees and lighting up an absolutely gorgeous morning. I had to wonder what amazing things I'd see along the way to town. Remember the shaggy yearling doe I wrote about several weeks ago? This morning she was back in the middle of the road staring the opposite way as I approached. Her udder was still full so she must be raising her baby and a few weeks of good grazing has put a layer of fat on those once skinny ribs. The shaggy hair has been replaced by a sleek and shiny reddish summer coat. She was truly a thing of beauty as she bounded across the ditch into the meadow.

Kevin was awake when I got to the hospital. When Margo heard us whispering she opened her eyes and smiled at me. I went straight to her side and kissed her and told her I loved her. Slowly but clearly she said, "I love you, too." I broke down and wept. At once she wrapped her arms tight around my neck and pulled my face down beside her. She kissed my cheek and told me not to cry because it would all be okay. That got Kevin and me bawling. What a tender moment and to think that a month ago, I was expecting her to go at any minute. This past month has been God's blessing and so precious to Margo and me as well as to our family and friends.

I managed an hour's sleep in the chair before the breakfast trays arrived. Margo drank her laxative but barely managed to swallow the five pills I gave her. She couldn't bring herself to eat a bite of breakfast. It's come time to respect the promise we made to each other many months ago. When she's too tired to fight any longer she can quit eating and fade away.

A nurse talked to me about switching the morphine from pills to receiving it through her intravenous line. This will keep her as comfortable as possible.

Ruth came and stayed with Margo while I went home to do some work on the farm. I had lunch at home and then went around by Lundar to attend to some business. Everywhere I turn I meet people who come to shake my hand and give me a hug as they wish me well. It totally blows me away and I'm so thankful to be living where I'm surrounded by wonderful people.

I got back to the hospital to find Margo sleeping. Ruth brought me up to date on all who had stopped in or phoned. Georgina was there visiting in the living room with Ruth and the three of us talked quietly for awhile. Margo woke up and was so happy to see Georgina. They had a bit of a chat before the ladies left Margo and me alone. I settled down for a half hour nap and then woke up to find Margo's brother, Art, and his lady friend in the next room. Art is going through severe health problems of his own but had still made the effort to come say goodbye to his beloved sister. Another brother, Paul, stopped in to see her and two of her sisters phoned.

Later Bubby peeked in and I took her straight to the bedside. Margo woke up and hugged Bubby close, calling her, "My dear, little daughter." I was amazed at how alert Margo was.

A short while after our company left, a miracle happened and I'm so very grateful I was there to witness it. I tidied the kitchenette and then went back to Margo. She opened her eyes and her face was radiant with a brilliant smile on her lips. She waved her hands for me to come closer as she said, "Come close. Come close." I bent over the bed and kissed her as she asked me, "Are you glad?" I asked her what she meant and she answered with another question, "Aren't you glad that I'm going there?" Her eyes were sparkling with happiness as she gazed deep into mine. I asked her if she meant Heaven and she replied, "Yes."

Then she told me she'd been hearing something. My heart started pounding in my ears as I asked what she'd heard. She said she could hear wonderful music. She looked away as if gazing at something right out through the wall. She really floored me when she said, "It is so beautiful here." There was a slight pause before she breathed, "So beautiful." I asked her again if she meant Heaven and she replied that she did. I told her I'd known she was going there and that she'd soon be happy and free from pain. She was so radiantly happy and that beautiful smile of hers was lighting up the room. Then she told me with another smile that she could even hear a little kitty cat there. I was shaking so hard by then that my legs could hardly hold me as I bent over the bed but I didn't dare move for fear of breaking the spell. That was about as far as the conversation went but it was enough for me to know that my prayers have been answered and it has strengthened my faith in life eternal.

I'm so afraid of leaving her side and having her pass on while I'm gone but late tonight I realized I didn't have my ulcer pills with me. I phoned Ruth and she got out of bed to sit with Margo. I wanted to do this Update while it was fresh in my mind.

Before I left the hospital, Donna phoned to say she arranged to take tomorrow off work. Brad got off a few hours early this evening and they are on the road already. Margo even woke up long enough to talk to her girl. We both look forward to seeing them in the morning.

I'm going to send this and then grab a couple hours sleep before heading back in to see my honey. I hope she waits for me. We're getting pretty darn good at this kissing game again after being kind of rusty for many years. Why do we let the important things slip while doing other mundane things like watching stupid television? Love, Dan"

The next day this wonderful message arrived from my sister.

"Dear Dan,

I'm praying you'll get your wish and your whole family will be gathered around you and Margo tomorrow. I cringe to think of the pain you're going through and yet your updates shine with the joy you're taking from these special days with your dear Margo. When Norm died it was so sudden that no one had time to say goodbye and it was the same with Mum. Which is worse, watching your loved one failing a bit more each day, or having them gone so fast that it's like a kick in the stomach? I hesitate to say it, because it's not a good thing to watch your wife die, but I do think you're lucky to have these special days with Margo. I doubt you've ever loved her more or felt as close to her as you do right now. I think it's wonderful that you can gather your children close and cry with them.

But, if indeed, Margo has had glimpses of Heaven, how can we feel sad for her? We can only feel sad for ourselves that she'll soon be leaving us. Please give all your kids a hug from me and tell them Auntie Kathy is praying for them and holding them close in my thoughts. Love to all, Kathy"

"Thursday morning Margo Update July 7, 2005

I just got home from the hospital after a 28 hour shift. Margo hung on for our anniversary and she's still breathing but barely. She wants to go on to a better place but her heart and body are resisting her leaving.

I was in there shortly after six o'clock yesterday morning and the first thing I noticed was that her breathing pattern had changed from slow and deep to fast and shallow. As the morning wore on she started to gasp for breath when she inhaled and moan as she breathed out. This is so hard to watch and listen to and I've begged her to let go and slide away on a golden cloud.

Donna, Brad and kids arrived before noon and Margo greeted all of them as well as she was able. Several times during the day, special friends stopped in to see her and I didn't have the heart to turn them away after they've all been so good to us.

This was the second day she hadn't been able to eat anything and she was receiving her morphine through the intravenous line. They've had to keep increasing the dosage as her pain worsened but she was still in agony when they tried to turn her in bed.

In the evening we finally had the family reunion, but it wasn't complete as Tracy was not able to take time from her new job to come out a day early. Our other three children gathered with me around her bed and we talked about old times, mainly about how those kids used to drive us half crazy. Every now and then there would be an outburst of laughter and Margo would open her eyes and smile at us. By late evening there was only Brad and I there. We talked to Margo and listened to her battle for her life. Why, oh why, can't she let go and fade away to find peace at last?

Dr. Steyn came in just before midnight and was distressed to see how Margo seemed to be suffering. She told us it wasn't acceptable and she ordered the morphine dosage increased again. Half an hour later she came back and checked her again and the laboured breathing was still the same. She turned her face to the corner of the room for a minute as she fought to regain control of her emotions. When she turned back to face us, there were tears glistening on her cheeks but she had a big smile on her lips.

She woke Margo and asked if she wanted something to make her sleep deeply and Margo whispered that she did. Elzette ordered something to make her sleep and then came back in half an hour. Margo was deeply asleep but the breathing hadn't changed. Elzette said it was all she could do but she was sure Margo wasn't suffering.

This morning Kevin was there early. We had our cry together before the nurses came to wash Margo. I'm so happy to have a wonderful support group and wonderful caregivers in the hospital. Those nurses have looked after all the personal details of Margo's hygiene without making her feel the least bit embarrassed. This is allowing her to pass from this life the same way she has lived it—a real lady. The nurses are astounded that even yesterday they got a tired thank you from her for every little kindness they performed.

After breakfast this morning, Margo managed to wake up long enough to kiss me and whisper goodbye but she was too weak to raise her arms. We gazed into each other's eyes for a long moment as if trying to read each other's mind and then I left after another tender kiss.

Please say another prayer for us. It's such a comfort to know we're surrounded by love from all sides. Dan"

FOURTEEN
Her Final Five Days

Within an hour of writing that last Margo Update I was back at her side. If I'd been hoping for more hugs and kisses, I was too late. Margo was in a coma, breathing softly and evenly with her face as serene as if she was enjoying a wonderful sleep. Perhaps she was! There were no signs of stress or pain and this was an immense relief to me after watching her struggle for every breath for the past few days. I knelt beside the bed and kissed her face while softly calling her name as if willing her back to me. There was no response except for a slight flutter of her eyelids every time I touched or spoke to her.

Drawing a chair close, I sat with my face beside her cheek and talked gently to her for what seemed like hours. The Do Not Disturb sign was on the door and we were left alone except for a nurse occasionally checking her vital signs.

Finally, exhausted, I sat in that comfortable recliner hoping to get some sleep. It wasn't meant to be. Trying to relax just made me wider awake. Every few minutes I'd look at Margo, hoping against futile hope that her eyes would be open and she'd be gazing at me with that special look of love I'd always taken for granted.

So many times I'd prayed for her to slip into a peaceful coma, one so deep she wouldn't feel pain or bitterness or fear. Now it had happened and all I could feel was regret and deep remorse. Why hadn't I stayed with her that morning, speaking to her until the final second when her last shred of consciousness slipped away as silently and fleetingly as if it had never existed at all?

Over the past few weeks I'd spent hours sitting in that chair reading a book, while Margo gazed vacantly at the television screen. How I regretted those wasted moments! I yearned with all my heart to be able to speak to her just once more and to hear her gentle voice answer.

Why had I spent so much time hunting the previous autumn? Why hadn't I realized that these were our last precious moments together that I was carelessly squandering? Day after day I'd awakened early and left her sleeping until she awoke alone in her empty bed. Why hadn't she scolded me and told me she was lonely without me? Surely I would've understood her need and done all that I could to please her. Why had I spent countless afternoons in the bush cutting firewood for the house, driving myself to exhaustion as if this was the most important thing in the world to me? I had enough cut wood to last for years so there was no need to be away from the house at all. Perhaps it was my way of coping with the situation and relieving stress and anxiety with physical exertion but now looking back, there was no excuse for my behaviour.

The afternoon passed in an agony of deep regrets until I finally fell into a fitful sleep. The evening brought visits from our children tip-toeing quietly into the room as if they were afraid they would wake their mother. It was my turn to be strong as I comforted first one and then another as the truth dawned on them, as it had on me earlier in the day. There would be no more chances for soft conversation. No more gentle hugs and warm kisses—only the slow rise and fall of her chest to assure us that she was still with us.

I spent the night in the chair at her side, afraid to even go for a quick walk down the hall in case she drew her last breath while I was away. I'd promised her I would be beside her until the end and I intended on keeping that promise.

At regular intervals a nurse would appear and I'd help her move Margo into a more comfortable position. There was never a change in that gentle breathing except for a slight quickening when I spoke to her and always the flutter of her eyelids which told me she was still responding to my voice.

At one point in the early hours before dawn it occurred to me that Margo's life had come full circle. She'd begun her life here on earth as a helpless infant with her every need being tended to by loved ones. Now, once more, she lay there like that helpless baby unable to do more than breathe.

Daylight arrived and Friday passed slowly with no change. My children arrived bringing me fresh clothes. I went down the hall to shower and shave before hurrying back to Margo's side almost afraid to look at her for fear she'd left me while I was gone.

That evening Tracy, Rick, and The Boy arrived. They had hurried out from the city as soon as they got off work. A short while later Rick took Dylan over to Steven's house while we had another family evening. This time our group was complete with Tracy and Ruth with us. We talked quietly and reminisced about their years spent growing up on the farm. Margo was unresponsive.

Alone again later, my thoughts turned once more to, *Why?* Why had so many of our friends been stricken with cancer? It seemed as if half the people we knew had either survived cancer or succumbed to it.

My gaze fell on a package of store-bought muffins Brad had brought me. They looked so appetizing in their fancy plastic packaging but they really weren't that good. I picked up the package and read the label. It was a shock to see they might've been concocted by the devil himself with a list of ingredients guaranteed to give them a shelf life for months even in the depths of Hell. There were eight different types of sodium listed, sodium this and sodium that, as well as many other scary sounding ingredients. What happened to the good old days when our mothers whipped up a batter of natural foods like flour, sugar and milk with maybe a handful of raisins for extra flavour? The finished product was so good the muffins were eaten within a few days or sometimes even before they had a chance to cool off.

We're living in a world where we want everything to be totally convenient. We choose products that are packed full of preservatives, no matter if some of those ingredients will surely cause us an early death if we eat enough of them. Step outside and take a big breath of fresh air. Well my friends, it's not fresh anymore, even living where I do far from the nearest city. Every breath we take is laden with chemicals or some sort of radiation guaranteed to bring on lung cancer in those of us who might already have a weakened immune system. Mankind must make some meaningful changes in this beautiful world or soon there won't be enough healthy individuals to care for the sick.

Totally exhausted I fell into a restless sleep which gradually deepened as the night wore on. Early Saturday morning I woke with a start to find Margo still breathing quietly. The day passed slowly with a continuing series of thunderstorms passing through the area until it seemed as though the whole country side would be flooded.

Darkness brought another vicious thunderstorm with huge raindrops pounding against the window. I held Margo's hand and stroked her hair as my mind wandered back to the early days of our life on the farm. An involuntary chuckle escaped my lips as I remembered my little wife angrily running around our old log house putting pots and pans under the leaks in the roof. How I wished I could turn back the hands of time and start over right there.

Midnight came and went but still the storm raged. Sleep refused to come as my mind continued to explore one thought after another. Even the distraction of television only caused me to feel more agitated. The subways in London had been bombed by terrorists and every television station was covering the disaster over and over again in order to corner their share of the ratings. I tried to fathom the reason for man's cruelty to man. What could cause man to blow up innocent victims as a form of revenge? What was the answer to all this senseless cruelty? Not retaliation.

The television brought reports of flooding in Manitoba and I was reminded of the story of Noah and his Ark. That had always been one of my favourite stories but it was also one that I had a very hard time believing. The simple mathematics involved in cramming that many animals into one boat, along with the feed required for that many days were a bit more than I could swallow.

There was also the story of Jonah and the whale which was another wonderful but unbelievable story. My time spent dressing salmon on a fishing boat had taught me to wear rubber gloves or risk stomach acid burning the skin right off my hands. How could anyone survive inside the belly of a whale and emerge unscathed three days later? It wasn't that I didn't want to believe; it was just that I was having a hard time believing.

Many years earlier, I'd gone to a funeral with the service conducted by Bill. During his sermon he spoke about how we sometimes struggle to interpret and understand what we're reading in the Bible. He told us about a weekly Bible study he'd attended with a small group of neighbours where there had been many arguments about which Bible stories were true and which ones weren't. Bob Kingsley put an end to the arguments when he told them that they had to make up their own minds and then stick to it. He said that it was either all true or not true at all. Those words made such an impression on me.

I finally relaxed enough to fall asleep in the chair for an hour or more only to be awakened with a start by another tremendous flash of lightning and rumble of thunder. It was raining harder than ever. I moved closer to Margo and found her condition unchanged.

Deciding I had to get out for a breath of fresh air, I walked down the hall and out the front door to a sheltered spot where I could watch the storm. Jagged streaks of lightning ripped through heavy black clouds, releasing torrents of rain that fell like a solid curtain to the already saturated earth, while deafening claps of thunder shook the building.

With most of southern Manitoba covered with water so that it looked more like ancient Lake Agassiz than fertile farmland, it wasn't hard to imagine what would happen if the rain continued for another forty days and nights. For the first time I found it easier to believe the story of Noah's Ark.

I watched the storm and prayed for God to take Margo to Him. I spoke softly to Margo as if she was beside me, asking her to just let go and drift away. She could hitch a ride on those towering clouds and powerful updrafts. By the time her spirit rose above the storm into the clear air above, she'd be halfway to Heaven.

Before I'd finished speaking, there was a sudden gust of wind, combined with a bright flash of light off to one side, along with the sort of rushing, rippling sound a large bird of prey makes when it enters a power dive. It lasted no more than a few seconds but it frightened me badly. My first thought was that Margo had died. I hurried back to her room to find her still breathing and looking exactly the same. I stroked her head and talked to her until the storm passed with the first light of dawn. It was a new day, Sunday, July 10, 2005.

By breakfast the rain had stopped and the weather report brought news of clearing skies with good weather for the coming week. Our kids arrived, bringing another change of clothes and an offer to sit with Margo while I went home for a sleep in a real bed. I refused the offer with thanks, opting instead for a shave and a quick shower in the tiny room down the hall. That afternoon a few friends dropped by from time to time, giving me a chance for quiet conversation in the adjoining sitting room. After keeping my thoughts bottled up in my head all night, the verbal volcano was in full force and my pent-up words tumbled out non-stop.

The long night that followed seemed never-ending. I sat in the chair and read until sleep overcame me. Sometimes I'd wake with a start when the book fell to the floor. I continued to speak softly to Margo and was rewarded by flickering eyelids and the quickening of her breath.

As daylight approached, I was nearing the end of the book I'd been reading. A few days earlier a nurse had lent this book to me and it had fascinated me. It was a story about a man who had fallen out with his father and had never forgiven him. It caused me to do a lot of thinking about the grudge I'd held against my own father for the severe beating he'd given me when I was a kid.

The story in the book had a happy ending with forgiveness and loving words from both parties. I closed the book and sighed, wishing I'd found the courage to forgive my dad while he was alive. Oh well, the story was fiction. Fiction lets the author select any ending that suits his or her fancy while in real life we're bound by restraints of stubborn pride and bitter anger.

Daylight arrived and brought a horde of noisy birds to the feeder outside the window. I wondered if Margo could hear them and if she was taking pleasure from the sounds as she always had. It was early Monday morning, July 11, 2005.

There was barely time to finish breakfast before my two sons arrived. It was so hard to see them go to their mum and speak to her as I'd been doing, as if the sound of our voices would bring her back. Once more they offered to sit with Margo and let me go home for awhile but again I opted for a shave and shower along with a change of clothes. I felt that it wouldn't be long before Margo left us. Her breathing was becoming softer as the hours went by and sometimes you had to watch closely in order to see that she was, indeed, still breathing.

Back in her room again I felt restless and asked the boys to sit with Margo while I went for a brisk walk to stretch my legs. They told me to stay away as long as I wanted as they wanted nothing more than to spend time beside their mum.

The skies were clearing as I left the hospital. I intended on walking a couple of miles but my walk was cut short when I reached the old Anglican Church on Railway Avenue. Unused for years, it had been converted to a museum and had served well in that capacity. As I passed the building I noticed the front door was open and a sign was set out inviting visitors to stop in. I decided at once to go inside to look at the old oak arm chair my dad had lent to the museum.

Dad had decided at one point to build a reclining chair using instructions he found in a magazine. He used oak lumber that he'd sawn on his own mill. The end result was a chair that was solid and strong but far from comfortable, mainly because he hadn't used much padding in the cushion. That didn't matter to him though and he often spent rainy afternoons reading and dozing in that chair. The chair was left behind when he moved to BC.

When we moved into our new house in 1974, the chair was left in the old house until we tore it down a few years later. Dad was in Manitoba visiting us that winter and he helped with the demolition, even though this must've been extremely painful for him. It wasn't long before he became somewhat depressed and decided to cut his holiday short. Before he left, he spoke to me one day, telling me he'd be honoured if I'd accept that old chair as a gift from him. Without giving his feelings any consideration I told him we had more than enough furniture in our new house and that I'd never found that chair to be very comfortable. He quickly turned his back on me for a moment before he replied that it was alright as he'd find some other home for it. Within a few days he'd made a deal with the museum board to lend it to them indefinitely.

Back in a corner of the museum, I found the chair looking much the same as it had when I was a little kid. I stood with one hand touching the chair's arm and was deeply ashamed to think about how my brusque reply must have hurt my dad.

Ignoring the sign that warned visitors not to handle the articles on display, I brushed a layer of dust off the chair and sat down. It wasn't all that uncomfortable now that the intervening years had put a layer of padding on my butt.

I leaned back and closed my eyes as my mind went back to the day I lay crying in the bush until after dark, my body aching from the beating I'd received at the hands of my father, while my mind formed a firm resolve never to forgive him. My thoughts did a fast-forward to the years when as a young father I'd been far too harsh with my own rambunctious children in an attempt to discipline them. It was well past time to be totally honest with myself. When I had spanked my kids more than they deserved, was I merely trying to discipline them? Or was I trying to get back at my own father for the one and only time he'd lost control of his emotions and punished me much more than I deserved?

As I matured, I became a much gentler person to the extent that our two youngest children often escaped with very little discipline. To my surprise, they never grew up to be juvenile delinquents. They are honest hard-working citizens like their older siblings and without the pain involved.

The older I got, the more I regretted how I'd treated those dear kids and apologized many times to them for my cruel actions. They told me they loved me and they forgave me. They even tried to make me feel better by saying they'd deserved it. Of course we all knew better than that.

I sat in that old chair while the tears came in a torrent that almost equalled Saturday night's downpour. How could I have been such a hypocrite to ask my own children for forgiveness when I was still holding a grudge against my own father? Through a heart that felt like it was breaking in two, the words came pouring out as I spoke to my dad, telling him that I forgave him at last. I also asked him to forgive me for holding that cruel grudge all these years. Last of all, I begged God to forgive me and I'm sure He heard my words that sad morning.

When I felt calm again I ran the few blocks to the hospital, anxious to see Margo again and not sure if she would still be alive when I arrived. I felt as though I'd been washed clean of my sins by my tears and genuine remorse.

My sons looked up, startled when I walked into the room with my face swollen from crying. I called them to me in the sitting room and explained where I'd been and what had happened. Once more I asked them to forgive me. Those dear sons will never realize how much it meant to me when they gathered me into their arms and cried with me. They told me there was nothing to forgive. They stayed with me until my tears stopped and then I urged them to go home and be with their families.

I sat beside Margo for a long time, telling her about my experience at the museum and letting her know that all was well between my dad and me. Again there was the flickering of her eyelids and a slight movement of her face toward me. I held my breath, hoping for some miracle that would bring her back even for the briefest moment but there was none. It didn't matter though as I was sure she'd heard and understood my words.

The afternoon brought bright sunshine. After lunch, I phoned Harriet Kaartinen to tell her how Margo was doing. I told her that I'd planned on having a walk to get some exercise but my plans hadn't worked out. She offered to come in after supper to sit with Margo while I went out for my walk. I told her I'd try to sleep for a few hours and would welcome the chance to get out in the evening.

Before I sat down for a nap, I went close beside Margo and talked to her again. It was almost unbelievable that she was dying right in front of my eyes. She was still warm and beautiful and serene.

I'd barely settled down in the chair when a nurse came to tell me that there were visitors at the desk. Looking down the hall, I saw our good friends from St. Laurent, Emile Lavallee and his wife, Marcelle. I called to them to sit with us for awhile and their presence turned out to be such a blessing. I told them Margo was almost ready to go and meet the Lord. They sat down on each side of her bed while I got busy making a pot of coffee. I could hear Marcelle speaking softly to Margo in her beautiful French accent.

Rejoining our friends beside Margo, I asked Marcelle if she'd talk to Margo in French because with all of the French ancestors in Margo's background, there might be a chance that she could hear and understand every word. To this day, I have no idea what Marcelle said to Margo but it brought tears of happiness to my eyes and peace to my heart. I can only imagine what it might've done for Margo but Emile, who had been watching closely, pointed out her flickering eyelids as proof that she could, indeed, still hear and understand.

I begged my friends to stay as long as possible and with their wonderful company, the time passed quickly. They finally had to leave and there was still time for that much needed nap before the supper tray arrived. For the first time in days, I ate with a hearty appetite.

Shortly after seven o'clock, Ruth arrived to spend a few minutes and then Harriet arrived to sit with Margo while I went out for my walk. The three of us chatted quietly before I stood up preparing to go. Before I left the room I took Margo's hand to give her a kiss goodbye. She didn't look any different than she had for the past few days but something made me decide not to leave her side after all. Somehow I knew she was only a few minutes from death.

I called to Harriet and Ruth to sit close beside Margo across the bed from me and to hold her hand and touch her arm. Several minutes passed as I held her hand in mine and watched her relax as her breathing slowed. The back of my hand was right over her heart and I could feel the beat getting slower and slower until it finally gave one last flutter and stopped. At the last moment she gave a tired sigh and turned her face toward me for the last time.

My sweetheart was gone and I felt immense relief that she was out of her pain. There would be time to feel deep sorrow and grief later but for now I was thankful her battle was over and that I'd kept my word by staying with her until the end.

I'd like to close this chapter with a few paragraphs from the eulogy I wrote and delivered at Margo's funeral on the following Saturday morning.

"Just like I promised you, honey, I stayed with you right until the very end when God chose to part us at last. After you left me, I stayed behind in that wonderful palliative care suite that had started to feel like home to me. I spent some time phoning family and friends while waiting for John to come pick up the tired body that you didn't need any more. When John arrived, I helped him move your body onto the gurney. We both knew you couldn't feel any pain but we were ever so gentle with you. I tucked in your little feet for the last time and then gently zipped up that soft bag until just your face was showing. I gave you a few more kisses from the kids and one bigger one from me. After telling you once more how much I love you, I closed the bag the rest of the way.

John graciously allowed me to help him wheel you out of the hospital all the way to the waiting van. I think our dear nurses were surprised to see the smile on my face but all I could feel right then was relief that your suffering was over at last.

It took only a few minutes to clear the room and make my way out to the car. The sun had just fallen behind the horizon, lighting up the fluffy little clouds with a radiant rose-coloured light. As I crossed the highway heading west I was greeted by an amazing sight. Thousands upon thousands of sea gulls were heading south to their roosting places on the shores of Swan Lake. It wasn't hard to imagine you flying with them on the first stage of your journey home."

FIFTEEN
Margo's Celebration of Life Service

My dearest Margo,

July 17, 2005

It seems strange to be writing a letter to you, less than a week after your death, but that is exactly what I'm going to do. It's going to be a fairly lengthy letter, something like the letters I used to write to you when I was a fisherman on the west coast in the sixties. That was back when you were a young stay-at-home mum. I still have the pictures you sent me that summer when Donna was a baby. In the picture I liked the best, you were sitting primly with the baby on your lap while you gave me that sexy look I used to find so irresistible.

Oh, who do I think I'm kidding here? I still find you totally irresistible even now after you've left me for your new life with our Lord in Heaven. My love and passion for you has grown so strong over the past few months until in our final days together it became almost unbearable.

Ever since your spirit left, you've been dragging me out of bed at all hours of the night to witness beautiful signs from Heaven. These special signs you've managed to send have given a whole new meaning to the word "supernatural" and have left our whole family gasping in awe and wonder.

I arrived home a few hours after you left us on Monday evening, where I joined Kevin, Nat and their kids, as well as Tracy and little Dylan. They wore a bunch of long, sad faces when I first walked in the door but within five minutes, my exuberance was catching and I had them all laughing. I've done plenty of crying since then, honey, but that is to be expected.

The true miracle came later when you sent the special sign that you'd made that marvellous journey Home okay. Our kids knew nothing of your promise to me so it came as quite a shock when Kevin and Nat came bursting into my computer room where Tracy and I were talking. They were yelling that we had to come out because Mum was sending us a special sign from Heaven.

We held hands and groped our way out in the dark to be greeted by a display of northern lights, the likes of which I'd only seen once before in my life. That was the night God showed me a special sign back in 1999 when I thought I was still a non-believer.

As we took our places out on the lawn, Kevin was yelling that he'd seen you waving at him from the lights. This was coming from the same young man who just weeks ago had told me he was concerned for my mental stability.

We stood and gazed up at the sky and marvelled in what we were seeing. A brilliant band of lights extended right from the eastern horizon in a high arc directly over our heads until they fell into the lake on the western horizon. The rest of the sky was perfectly clear and so dark, the stars looked like diamonds. Part way up from the eastern horizon, the band broadened to about twice its width and in that spot the lights were revolving slowly, with just a hint of the brilliant colours that we sometimes see in the dead of winter. Kevin spoke in wonder about how he'd seen you waving at him from that spot and I replied that I'd always thought of northern lights as angel wings. He insisted that it was his mum and once more I replied that you are now one of God's angels and safe with Him in Heaven.

Nat pointed out a brilliant star directly over our heads and said she was sure that it was Mum. As the band of light kept passing back and forth over that star, a perfectly round hole would appear in the lights with that brilliant star directly centered in the hole. I had to agree that it was you all right.

About one o'clock in the morning, I tore myself away and went to bed. My sleep was deep and dreamless until three o'clock when I woke suddenly with a start. I decided to go outside to see what was happening with those lights.

Out on our deck, I looked up and saw the lights were still there, though not as brilliant as before. As I looked east over our house roof, I saw what looked like a bright star just to the south of the band of light. It couldn't have been a star because it was moving across the sky and it was moving much too slowly to have been a meteor. It occurred to me that I might be seeing a man-made satellite for about the fourth time in my life. I watched that bright golden globe moving straight toward the angel cloud in the middle of that band of light and wondered whether the satellite would pass in front of or behind the lights. It took maybe twenty seconds for the craft to reach the lights and disappear and I estimated that it might take about four seconds to reappear out the other side. I was counting the seconds and started saying, "Right about now, it should reappear." Seconds ticked by and still it never showed. I held my breath until it finally dawned on me that I was never going to see that golden craft again. When the truth finally dawned in my thick skull, I burst into tears and fell to my knees out there on the deck. I stayed there for another ten minutes and that craft never showed again. I knew it was you, dear Margo, giving me one last glimpse of your special beauty. I have only God as my witness as to what I saw but I know that should be more than enough.

I didn't sleep much the rest of the night and was in the kitchen when Kevin and Nat and their kids came in from their motor home. They still looked so excited and I couldn't wait to see the look on their faces when I told them what I'd seen after they went to bed. They were amazed with my story but they had an even better tale to tell me.

After I'd left them alone in the night, they sat spellbound for about an hour, unwilling to break the magic spell. The lights grew brighter until suddenly the pulsing cloud of angel wings gradually moved westward across the skies until it was directly north of them. The lights kept swirling around until a huge, perfectly shaped heart appeared with the point hanging straight down with the tip touching the roof of our house where I was sleeping.

This was too much for me and I burst into tears again, totally startling the little kids. When I'd recovered my composure, Kevin and Nat told me they'd watched as the heart shape faded away to the east to return to the band of lights. They'd been so spooked they went straight to their beds feeling they'd seen quite enough miracles for one night.

You did it, my sweet one. You sent us a sign that was so beautiful and so powerful that it was almost beyond imagining. How God must love you and how special you must be for Him to allow us that wonderful glimpse of what awaits us when we leave this earth, if only we can relax and put our complete faith in Him.

For the past few months I've been feeling I didn't want to continue to live in our house on my own after you left me and I'd been considering all sorts of options. Now, in your wisdom, you've shown me there is no reason for me to feel alone here. There is no place on earth that I love as much as our farm and now I'll be content to stay here with you watching over me.

On Tuesday, we all caught our breath and rested up a bit. I worked on your eulogy all afternoon and came close to completing it. I felt so at peace and completely calm for the first time in months. The words flowed smoothly as I poured out my love for you in that message.

That evening all of us were in bed by eleven o'clock. I fell asleep at once but woke up after three hours of deep sleep and found myself wanting to get outside on our deck to pray to God. I also wanted to talk to you while I stood in the same spot where I got my last glimpse of that golden craft in the sky the previous night.

It was a warm and calm night with hordes of ravenous mosquitoes swarming over every inch of exposed flesh. I decided this was going to have to be one very quick prayer with my eyes turned toward the night sky. I spent maybe thirty seconds slapping mosquitoes before I got serious with my talk with God. Strangely enough, as soon as I started speaking, the mosquitoes quit stinging me, or maybe I stopped feeling them.

I stood speaking to Him with my eyes wide open and I wasn't all that surprised when I noticed a light in the sky. This time it was a high-flying jet up there in the frosty air at thirty thousand feet heading east. Just to be on the safe side, I watched it go until I was afraid to blink for fear of not being able to spot it again. Just when it faded out of sight, a brilliant, golden light appeared in the sky just out past our television mast, in almost exactly the same spot where I last lost sight of that golden craft.

The light was only there for maybe half a second before it disappeared, only to reappear again about a second later. This time the light was still bright but not as bright as that first brilliant flash. It was in exactly the same place, as if someone was flicking the switch on a giant spotlight way out there in the heavens. Before I could even gasp, the light disappeared once more and then flashed back on again. Again it went out and then flashed for the fourth time. Each time the light came back on, the flash was dimmer. After the fourth time, it never appeared again.

I stood there for a few more minutes until the mosquitoes descended on me in force. Making my escape into the house, I came straight through to my computer-music room chapel. I sat in my big, leather chair and tried to make sense of what I'd seen.

There is probably a logical explanation for this sighting. It could've been the sunlight reflecting off the window of a space station way out there beyond the reach of the earth's gravity. Is it possible the stress of the last few months has caused me to lose my mind? I don't think so and I think our kids will agree that I'm acting in more than a rational manner. Not only that, but Kevin, Nat and Tracy all saw the same thing I did on the night you left us.

Now, here's another strange twist. Harriet was with us holding your hand when you died and that sent the poor gal into hysterics. I was able to calm her with my eyes and a few gentle words in order to let your passing be the peaceful one we'd hoped for.

Yesterday I phoned her to see how she was making out. After we chatted for a few minutes I started telling her the story about the northern lights the kids and I had seen that night. She got very excited and interrupted my story to tell me she, too, had been unable to sleep that night, so she got out of bed at two in the morning and went to the window in her living room looking south toward our farm. She saw that brilliant display of lights tumbling around like angel wings just like we had. So far I haven't met anyone else who saw them. Keep in mind also that this was at exactly the same moment when Kevin and Nat saw that angel heart descend until it was touching our house where I was sleeping so peacefully. Kind of boggles the mind, doesn't it?

I have to ask myself why God is spending so much time with me, a horrible sinner who turned my back on Him for all those lonely years. I know the answer to that will be revealed to me in time and I'm not afraid of what the future will bring. Just think about what you had to face and what your final reward turned out to be. I can only hope to be half as lucky.

Once back in bed I fell asleep and never heard a thing until the kids all woke up and came in for breakfast. That really turned out to be a quiet day and I was able to finish working on your eulogy. Nat looked after Dylan while Tracy helped me make a tape of the eulogy to use at the service. Tracy also suggested I take the copy of the eulogy to the service in case something went wrong with the tape. Donna and her family came back from Selkirk where they'd been visiting with her Uncle Art for a few nights.

Thursday turned out to be wonderful from start to finish. It was so nice to have the whole family gathered around me and I was very sure you were smiling down on us as we went about making plans for your service, which would be held on Saturday. We also worked around the yard in the morning cleaning up the clutter I hadn't had time to attend to earlier.

Later, we went up to the cemetery and cut the grass so we could pick out a spot for your final resting place. The yard was quite wet after all the rain we've been getting but we got the grass cut and trimmed between the grave sites. It did my heart good to see those strong, young grandchildren running around doing their best to help. I'm sure you would've been properly horrified to see Becky and me playing with a baby garter snake.

Back at the house, I found a beautiful message from a special friend, Hulda.

"Subject: Life carries on. July 14, 3:31 PM

Hi Danny,

I've planned on sending an email for a long time but have had trouble putting my thoughts and feelings on paper. Margo was such a special person, and so strong right to the end. Many things crossed my mind as I kept her company in the hospital... our fun Sequence games (we could've beaten you guys every time if we'd really wanted to!)... watching the two of you dancing... listening to her talk about the kids and grandkids... she was so proud of each and every one. I vividly remember the night I spent with her in the hospital, trying to ease her pain as Steven was making his entry into the world (he seemed to be in no hurry) and also holding him on the way to Winnipeg when he was only a few days old and so sick. You and Margo sat in the back seat looking so worried. I think Sandy Miller drove us.

Now that Margo's gone we'll miss her so much... her happy smile, cheery phone calls to chat... but although she's not with us in person, she's not very far away... watching us from a better place and waiting for us to catch up.

Danny, these dark days of grief will gradually brighten over time and we'll hang on to happy thoughts and look forward to better times. Our loved ones are never really gone. They're there in the dew drops, the rustling leaves and starlight. I saw this in a magazine and thought it fit.

"I said a little prayer for you and I know God must have heard

Because I felt it in my heart, Although He said not a word."

At times like these we all look at our own mortality and can only hope we follow Margo's example of dignity and courage, knowing the end was near. Reflecting on what values are truly important in this life... at these crossroads our hearts go out to you and your family. Take care, Danny. Love, Albert and Hulda"

Well, my sweetie, you can be darn sure I shed some tears over that note but they were happy tears too.

That evening was so special, when we all went down to the beach where I could spend some real quality time with our grandkids. Steven and his family couldn't make it but I had five happy kids gathered around me in the water. Nat took pictures of those cheeky kids splashing old Poppa. It's so wonderful to look at those pictures and see the joy on our faces. I really felt like I was starting to come alive again for the first time in a very long while. Back at home, we hooked the camera to my computer and made a special Margo's Memories file. Nat had been busy with their new camera for the past few days and there are some very special pictures saved in that file.

On Friday morning we had a conference and decided we should all go our separate ways and pick wildflowers for your service. We'd requested donations might be given to the Cancer Society rather than people bringing flowers. We knew how much you like flowers though and, since all the recent rains had the countryside bursting with blossoms of all descriptions, we intended to pick your very own personal bouquets.

That afternoon I took Dylan with me on the quad as I drove west out to the lake to see what kinds of flowers we could find. He loves riding on that big bike and he kept turning his head to chat with me. I was gazing at his face and trying to decide which one of us he resembles. Sometimes I feel as if I can see your dad, his Great Grandpa Nordal looking back at me. The truth is he just looks like himself, our little Dill Pickle, The Boy himself. He's a lot like you, very good natured but he also has that stubborn streak that served you so well.

You are a mixture of races, honey, and that mixture made you so proud. Is it your native blood, the blood from your Icelandic father, or maybe the Scottish or French blood that made you such a unique and wonderful creature? Whatever it was, that special blend of blood made you a very stubborn woman who was more than my match on many an occasion.

I always thought I was brave and strong but you taught me more than I could've ever known about how to be strong. When I tried to tell you how brave you were, you said you didn't have any choice but to be brave. I'm here to tell you today, my sweetie, you did have a choice. You could've chosen to give up and die in a few months, but instead you chose to fight for your life. As a result, you managed to live long enough to see two more fine grandsons born and to hold them in your loving arms as you watched them grow strong and ever so beautiful.

Just think about those grandkids, honey, eight of them—five strong boys and three beautiful girls. Each and every one of them has your blood running through their veins. How can I ever be sad again when all I have to do is look into their eyes and see you looking back at me? Our love for each other has started a legacy that will live on forever.

Dylan and I reached the special spot where Brown Eyed Susan flowers grow and in a few minutes we'd picked more than enough. We then took a few minutes to roll around in the tall grass, just looking up at the sunshine and the beautiful blue skies.

On the way home we passed over a rough spot. As I tightened my grip on that precious boy, I could feel his little heart beating under my hand. It reminded me of the way I felt your last few heartbeats under my hand as you left me. Tears fell again as I realized although your heart may be stilled, it will continue to beat on through the years in the breasts of your children and grandchildren and all the future generations to come. You will never be forgotten, my dear one.

Back home again I started mowing the lawns with Dylan's help, of course. As we passed the front of the house, I realized there were two large blossoms on our big rose bush. This is the bush given to us by Therese and Paul at our big party and dance in town a little over three years ago. So much has changed since that day. Both you and Therese are gone now, taken from your loved ones by that devil cancer.

The strange thing is that this bush has never bloomed until now so it came as a shock to see those twin roses both growing off one slender branch. One was just coming into its full bloom and the other was a bit past its prime and starting to wilt a bit. Yielding to a sudden impulse, I carefully snipped off the branch and brought the fragrant blossoms into the house where they were placed in a beautiful vase to take to your service the following day.

Bill and Joanne came for supper that evening and we barbecued steaks on the deck. Bill and I sang several songs out on the deck with one of your huge, wildflower bouquets in front of us. Daniel did some taping with his camcorder.

It was getting late by the time we finished supper. Phyllis Penny arrived to take the family portrait we'd been wanting for so long. It didn't seem the same without you there with me, honey, and there were some long, sad faces in that portrait. The mosquitoes were feasting on us, making it hard to hold still long enough to get several shots. We were all tired from the long day so it was lights out early that night.

I was up early the next morning looking after all the details needed to make this remembrance service the special one that you deserved.

By eight-thirty, I was at the big hall in town in time to meet John Grey, our wonderful funeral director. He'd looked after so many details in order to make things as easy as possible and he'd insisted we meet at least an hour and a half ahead of time. We carried in the buckets of flowers and set up the memory board with all our special pictures. As we worked, some of our friends started to arrive early.

John and I tried out the tape of your eulogy and that's when things started to go wrong. My tape that had sounded so good at home came out of his big machine all garbled and muffled sounding. We tried everything we could to get it right but I could see it was impossible. The acoustics are bad in that hall at the best of times and this set-up obviously wasn't going to work. I was getting extremely stressed until I remembered I'd brought the written copy with me. There was no reason in the world why I couldn't stand up at the mike and read from the script.

When I told John my plan, he looked at me as if I'd lost my mind. I assured him it would all be okay. If I couldn't get through it, I'd pass the script to Bill who would be doing the service.

There was still enough time to check the special poem I'd written for you which had been printed on the funeral card.

My Poem for Sweet Margo

My Dear One

June 4, 2005

I sit here close beside you, gazing at your winsome face.
The lines of pain are gone now, washed clean by our Lord's Grace.
You have fought a painful battle, with a strong heart pure and free.
We have been praying for a miracle, but a cure was not to be.

You have lived a life of purpose, raising four kids strong and true.
The love with which you raised them, has now returned to you.
I watch your slender body, looking for a sign of life,
But I cannot wish you back, dear; you have been an awesome wife.

I wish we'd found God sooner, just think what might have been.
Let's just say better late than never, and you've reached God's final scene.
I touch you, my sleeping beauty, as you lay so soft and still.
Now find a place for the two of us, in that home beyond the hill.

By now the front door was open, the hall was filling and I still wasn't dressed in my good clothes. I'd arranged to change my clothes at Steven's house but there was barely time to make the change and get back to the hall. When I arrived, I was astounded at all the cars which were parked on both sides of the street and all the way down the highway. Our friends were arriving full force to pay their last respects. Seeing their sad faces, it was all I could do to hold back my tears as I joined the rest of the family in the rotunda.

The next half hour went by in a blur as I talked to members of our families who were arriving until there was a huge crowd. Tracy took me down the hall to introduce me to a group of her friends from work, eleven of them in all. I kept a smile on my face but there were hot tears trickling past the lump in my throat.

Ten o'clock came and went and still the crowd kept pouring in. John came to tell me there was still a long line-up outside and they wanted to get everyone seated before starting the service. By then I was so overjoyed and overwhelmed, I told him to take all the time he needed.

We were running half an hour late when we were finally led into the hall. As I'd requested, I was carrying the urn that contained your ashes, with Donna walking beside me. I felt as though you were walking right along beside me too, honey, and that made me feel so strong and calm.

I hope you were proud of us as we took part in this last show of respect for you. Your grandkids, Kristen and Steven, went up and sang a beautiful little song. Natalie read a special poem that had been written by her grandmother. Bill read the scriptures I'd requested—our favourite passages from the book of Job. He also read your obituary before speaking a few special words to our family.

Then it was my turn and I went forward eagerly, with no fear or nervousness. I was so anxious to tell our friends all about you and our wonderful life together. I didn't just talk about the good times. I also told them about the bad times when I was drinking too much and not being the kind of husband you deserved.

I'm so happy now that the sound system gave us trouble, forcing us to change our plans. Several times I paused in my reading to say a few words that weren't in the script and it was totally amazing to see most of the ladies holding tissues to their eyes. Even more amazing was the sight of strong men sitting there with tears pouring down their faces. I know I went on much too long, my sweet one, but when I was finished I'd said everything that was in my heart.

Then it was time for our walk out of that big hall. It was so hard to look at our kids, honey, and see their grief flowing. God helped me to remain strong and be there for them when they needed me. We stood outside in the bright sunshine for several minutes before joining the crowd of friends in the luncheon hall upstairs. The place was packed and I knew if I sat at the family table, I'd only be able to see a few of our friends. I went up and down the rows of tables receiving hugs and kisses from all. It wasn't just the ladies hugging me either; I've never been hugged by so many men in all my life.

By the time we finally left the hall, most of the crowd had dispersed, leaving a small cavalcade to drive out to the Scotch Bay All Saints Cemetery. We all gathered at the gate and then proceeded across the grounds to the plot we'd picked out for you. Your two strong sons had dug the tiny grave for the urn containing your ashes. I walked slowly, carrying the urn, stopping with friends who wanted to touch it while they said their last special goodbyes. Unbeknownst to me, Nat was standing off to one side taking pictures that will be treasured for the rest of my life tucked away in your special Margo's Memories folder in my computer. The urn was placed on top of the grave and Bill started the service.

At the very moment he started speaking, a huge, green dragonfly came floating down out of the sky and landed on Daniel's sleeve, just above his left elbow. I was completely mesmerized at the beauty of this creature and was holding my breath as Becky reached out to touch it. That beautiful creation of God continued to sit there until Bill had concluded his short sermon. As he picked up his hymn book and started to sing, the dragonfly left Dan's arm and flew up into the sky. It looked as though it had set its sights on the white, fluffy clouds which were so much like the towering clouds that had swept me toward the heavens on that windy day back in 1983 when I was learning to fly a plane.

I kept my gaze fixed on the dragonfly until it was almost out of sight. Suddenly it turned and started down again, swooping and gliding in a descent which ended up in a steep, spiral dive toward the earth. At the last second it levelled its wings and swooped in over Bill's shoulder to make a skillful landing on the very tip of the hymn book that Bill was holding. It sat there facing him with its front legs reaching out toward him until he finished singing. As he closed the book, it flew upwards again. This time it never faltered, just kept going straight up until it faded out of sight. It wasn't hard for me to imagine your spirit hitching a ride on that big, green dragonfly all the way to Heaven. That must've been some wild ride, honey, and I can only guess at the joy you must have felt when you arrived at your marvellous new home.

Then it was time for close hugs and gentle words of comfort. I felt completely drained. Once most of our friends had left, our sons covered your grave and the huge bouquets of flowers were put in place. As the vase containing the twin roses was placed on the mound of fresh earth, the smaller, wilted flower fell from the branch. Donna and Brad had to leave for home right after the service but happily, Kevin and his family are staying a few more days.

You must've taken pity on me last night after seeing how tired I was after all the excitement was over with yesterday and this morning you let me sleep in a couple of extra hours. When I awoke, I felt no urge to even look out the window, just an overwhelming desire to come straight into my computer-music room chapel and sit in my big leather chair. I said a little prayer and then as my eyes opened, my gaze fell on the little clock on the front of the computer. The tiny hands told me it was ten to four in the morning and I felt an exquisite jolt of happiness in my heart as I realized this was another little sign from you.

This special time of day has become almost sacred to us over the many long years we've been together. Remember how we were sitting in my old car in your parent's yard when you gave me that first little kiss on the lips? I could've held you in my arms forever but you pulled back and asked me what time it was. I told you it was ten to four and that answer was the beginning of a little game we never tired of playing. We couldn't begin to count the many times that one of us would ask the other what time it was and the answer would come back "ten to four". That reply would bring a blush to your pretty cheeks and a warm surge of blood to my heart. We'd burst into laughter that would fill the space around us, swirling and twisting and turning in the air until the sparkling, silvery notes of our laughter were joined together as closely as our very souls have become over these many happy years together.

Today was a rather cloudy, cool day but we decided to brave the elements and take the kids for a boat ride down to Gull Island out in the middle of Lily Bay. As we neared the island, the waves became so large I had to go around the point and land on the east side which was sheltered from the wind. As we approached the shore, hundreds of birds of all descriptions flew up. There were the usual sea gulls and terns as well as large, ungainly pelicans and black cormorants. For the first time in my life I saw Blue Herons nesting high in the tree tops. It was a true bird-watchers paradise. We stepped carefully and spoke quietly to avoid disturbing the beautiful creatures. The camera was put to good use as we took turns taking pictures for your file. None of us really wanted to leave but the wind was cold and there was a rain storm approaching from the west. We launched the boat and hurried back to the warmth of our snug house.

This evening I watched a slide show of the pictures we'd taken and for some reason, all my attention seemed to be focused on one rather unremarkable photo of a young Blue Heron crouched in a nest near the top of a dead maple tree. There was something about that photo that kept drawing me back, time after time, until I realized perhaps it was the unusual cloud formation that was holding my attention. Like a little kid trying to compose pictures while gazing up at the clouds, I tried turning my head from side to side until suddenly an image jumped right out at me. Dead center in the middle of the frame was one dark, little cloud surrounded by towering white clouds. That little cloud was in the shape of an angel, complete with wings and halo, looking down at the bird in the nest.

You can be very sure I spent the next while just looking at that picture and crying but, once I became calm, I felt safe and secure once more. My darling, I miss you more than mere words can express but I need never feel alone again. If the loneliness threatens to overwhelm me, I can come into my little chapel and look at that special picture and know in my heart you are near me.

Goodbye, honey. I can feel the tears coming again but I just have to tell you once more that I love you. Keep a spot open there for me. I give you my word I will continue to live the kind of life that will allow me that special privilege, if God is willing.

I'll see you again some day, my dearest one. Goodbye and God bless you. Your husband, Dan"

AFTERWORD

November 27, 2007

Life is slowly changing from surreal and unnatural to happiness and normality.

When I first found myself alone, I was too exhausted emotionally and physically to feel anything other than intense relief that Margo had escaped her world of pain and suffering. Throwing myself into work to secure the next winter's feed for my cattle numbed my mind and the physical activity tired my body allowing blessed sleep to carry me away each night. During this transition period, I was unable to weep for Margo. It seemed my well of tears had run dry.

Here are a few paragraphs from my journal dated July 26, 2005.

"Things are gradually returning to normal around here. I've been busy from dawn to dusk in the hayfield and that has been wonderful therapy for me and for Ruth who has been here for me from start to finish.

Yesterday I caught myself singing a happy song as I went about my work and realized that my mood is brightening. It's like I'm awakening from a very long dream that alternates from sadness to despair to hope. I'm almost ashamed to admit it but I feel immense relief that Margo is at peace.

The burden for her care has shifted from me to one with shoulders much broader than mine.

I'm shocked to see how I've let things slip around the farm for the past couple of years. Fences are sagging, buildings need paint and the whole place needs a face-lift. It wasn't that I didn't have time to do this work. It's just that I had no interest. I felt I was slowly dying alongside Margo. I had no desire to continue living in our house without her beside me.

From the moment she sent me that sign from Heaven a few hours after she passed on, my whole outlook changed. I can feel Margo's spirit with me here on our beloved farm and especially in our house where we spent so many happy years.

Looking back at the events of the past two and one half years, I can see a pattern that has slowly led me to this present moment. My life has been drastically changed but I will survive and be happy again."

One day in late August, I found myself on my way to the annual corn roast which Les and Margaret Hayward had been putting on without fail for many years. Margo and I had seldom missed this event and, at the urging of several friends, I decided to break out of my rut and attend.

A mile from home I was driving along the country road enjoying the sunshine and blue sky when a small graceful hawk swooped down until it was keeping pace with the car, no more than fifteen feet from my window. Mesmerized, I watched it turn its head until its beautiful dark eyes were looking into my very soul. Our gaze remained locked on each other for several seconds until it suddenly veered away toward the heavens. Shaken, I stopped the car and burst into tears. I was convinced Margo had sent that beautiful bird to let me know she was doing okay and was more than happy in her new home.

On a bitter cold afternoon in October, Georgina and Raymond came for a visit which lasted until after dark. With the temperature near freezing, they bundled up warm and had just left the house when Raymond came dashing back in exclaiming loudly, "Dan, there's a big dragonfly on your steps and it's still alive."

The beautiful creature was there alright, so stiff with cold it could barely move a limb but it was very much alive. The three of us gazed at each other in amazement until I gently picked it up and carried it into the house to warm up. My friends were totally shocked but I was quite calm as I viewed my little house guest as just another sign from Margo. Later, the dragonfly warmed up enough to leave the arm of my chair and fly down into my computer room. I found it sitting on my computer and was writing an email about it when it flew out of the room never to be seen again.

Two years have passed since that cold night and there have been no more signs from Margo as much as I would like to have them. Sometimes late at night I walk out on the deck and gaze at the star-lit sky hoping to see more signs from Heaven. So far they haven't come and I have to content myself with listening to the sound of the night breeze in the treetops. At times it seems I can hear her gentle laughter and smell her special fragrance on the wind.

I've learned from my own experience that life does go on after the loss of a loved one. For the past six months I've been dating my beautiful lady friend, Fran, who for the past year and a half has been fighting her own battle with ovarian cancer. She has gone through major surgery and fifteen sessions of chemotherapy and is now doing very well. Her sense of humour and fighting spirit remind me so much of Margo and I pray we have many years together to enjoy our special passion for dancing with our friends.

Fran is one of the lucky ones. It seems as if every week we hear about another friend fighting for survival against that cursed cancer. Some of them will win their fight but others won't be as fortunate.

My faith in God has carried me through some very rough moments and my faith will stay with me until the day I die. I'm sure in my heart Margo made the journey safely to Heaven and if God is willing I will see her again some day.

LIFE WITH FRAN

Almost two years have passed since I worked on this manuscript. The editing was long since done and there was nothing preventing me from heading to the printer's shop. The truth was that I wanted to have the 'all clear' from Fran's doctors before finishing the book. We both wanted to say the treatments had worked and that she was in remission. Life doesn't always go the way we wish. On August 13, 2009, after a courageous three year battle with ovarian cancer, Fran passed on to a gentler world than the one she had known. I would like to tell you about the two wonderful years Fran and I spent together.

On a hot Saturday evening in August 2002, Margo and I attended a dance in Ashern, Manitoba. I had worked late in the hayfield and the dance was in full swing when we arrived. The hall seemed to be overflowing but a nice looking blonde lady invited us to join her and her friends at their table. She introduced herself as Fran Pettit and told us she lived across the lake from us near the town of Alonsa. She and Margo formed an instant bond when Fran commented on how much she loved Margo's shoes. Margo was pleased with the comment but, honest as always, replied, "Thanks. I bought them last week at Value Village." Fran laughed and said, "Oh, you're my kind of gal." Just like that, we had made a new friend.

It didn't take long to find out that Fran was a wonderful dancer and a good conversationalist. By the end of the evening, she and Margo had exchanged phone numbers and promised to stay in touch. To this day, her number is listed in Margo's careful hand writing in my address book. We couldn't remember her last name so it's listed under F as Fran from Alonsa.

We continued to see more of Fran at various dances that fall and gradually got to know her better. She was a widow having lost her husband, Fred, to cancer in 1993. After he died, Fran made her living by working at various jobs, even running her own used furniture store and pawn shop right beside her house along the highway south of Alonsa. She was the mother of two grown sons, Todd and Troy, and grandma to Brittany.

That winter Margo was diagnosed with cancer and started her three year battle with that cruel disease. We continued to go dancing whenever she felt up to it and in late fall of 2004 we drove to a dance at Kinosota. Margo's health was failing and we wanted to take in at least one dance in the hall that Fran had told us about. Fran was there that night looking elegant as usual and was so happy to see us. She told us that she lived just a few miles south of there and that her late husband was buried in the little Anglican graveyard just half a mile from the hall. She said in a matter of fact manner that her name was on the headstone with Fred and all that was left to be done was to add the date of her death whenever that happened.

The following summer Margo died and as I related, a large service was held in our town hall. Fran came to pay her last respects to Margo and I remember chatting with her in a hallway for several minutes. I was in such a daze by then I have no recollection of what we talked about.

In spring of the following year I heard that Fran had been diagnosed with ovarian cancer and was taking chemo treatments to prepare her for surgery. That summer I heard from friends that Fran was continuing to go to dances in between chemo treatments that were continued after her surgery.

Late in the fall of 2006 I drove to a dance and supper at Oak Point. I was hoping Fran would be there so I could dance with her and offer words of encouragement. Shortly after the dance started, Fran walked in looking so cute in a white wig that suited her to perfection. I was a bit disappointed that she had a man with her but after she introduced me to Herb Benson, I was pleased to see that she had such a nice guy for a friend. I had to wait to dance with her as it seemed like all her friends wanted a turn that night. By the time we danced a slow waltz, she was flushed and shaking a bit and the sweat was running out from under her wig. She was still game to go though and I never saw her turn down an offer to dance, even the fast ones.

Not long before the evening ended, I felt a little touch on my shoulder and turned around to see Fran smiling at me as she asked me to dance. She told me that she never asked men to dance and only did so now because she thought I wasn't going to ask her again. I told her I didn't want her to overdo herself since it had been just a few days since her last chemo. While we danced she said she was determined not to let the cancer slow her down in any way.

On January 17, 2007 I sold the last of my cows and subsequently found myself with nothing to do. For a week I slept in every morning, telling myself how wonderful it was to be retired. By the end of the month, time was hanging heavy on my hands and I accepted a temporary job at a neighbouring ranch helping to calve out over 900 cows. It was wonderful to be doing something I loved and the best part was sleeping undisturbed through the night in my own home.

One evening I called Fran to see how her treatments were going. She said her doctors wanted her to continue treatment until her cancer markers were down to a certain level. When I asked her if she was still dancing, she said, "Oh, yes, we go almost every weekend if the weather is suitable." When I asked her if she meant herself and Herb, she said, "Well, sometimes it's just Herb and me but if Marilyn isn't working, she goes too." I asked her who Marilyn was and she told me Marilyn was Herb's wife. My heart started hammering in my chest as I explained that I had thought she and Herb were partners. Laughing, she told me they were just old friends who had teamed up to save gas on the long drives to dances.

Late in March, I phoned Fran again and let her know there was going to be a dance held in Ashern the following weekend. I would be leaving early the morning after the dance to drive to BC to visit family. When I asked her if she was coming to the dance, she said, "I don't know. Are you going to be there?" I replied that if she was going to be there, I would too. She said she would make plans to come. We met at the dance and danced a lot together and it felt so good to be in her company. There was something in the way she looked at me that told me she was happy to be there with me as well.

By early May, I was back from the coast and Fran and I began meeting at dances almost every weekend. Being together felt right but we both had secret reservations about whether we should be letting things become serious when the state of her health was doubtful.

I finally screwed up my courage and asked her if she would like to come to my farm for Sunday lunch. She was out to the farm early and we spent a pleasant morning talking. It was a warm summer day and the sweat was soon running out from under her wig. I suggested she just take the darn thing off until she was ready to go home. She looked at me as if I had lost my mind but I soon convinced her that I was serious. She took her wig off and I kissed the top of her bald head while giving her a hug.

For the first time, we were alone and able to have a serious discussion about our future. Fran confessed she didn't think it was fair for me to become closely involved with another woman who was ill with cancer. That was exactly what my brain had been telling me, while my heart was telling me that here was a woman with whom I would gladly spend the rest of my life. We finally agreed that there are no guaranties in life and maybe we should grab our chances while we were able.

I had fish fillets thawed out for lunch and she taught me how to make fish gravy, something I had never even heard of. That afternoon we went for a long drive around the farm on my quad. I was quick to warn her that some of the land was a bit rough so she had better hold on to me. She was just as quick to wrap her long arms around me and hold me tight. A couple of times I felt her place her face gently against my back and this caused my heart to race almost out of control. Oh, it was love alright but neither of us dared to speak of it at that point.

All too soon the sun was setting and it was time for her to begin the long drive up north around the Narrows and back south to her house which stood just twenty-three miles west of my own house—so near and yet so far.

It wasn't long before we committed ourselves to each other and started spending most of our time together, either at my house or hers. I went with her several times to the Cancer Clinic in Winnipeg and sat with her while she had treatments. It gave me a sick feeling to watch her go through the same thing my dear Margo had gone through just a few years earlier.

It was amazing to learn that Fran had been a volunteer driver for Central Plains Cancer Care for several years. Even after she started her own chemo treatments, she had continued to drive sick people from the Alonsa area to Winnipeg for the cancer treatments and back home the same day. That would be an arduous routine for someone who is completely healthy and she had managed to continue doing so while very ill herself.

Fran sold her house and home quarter and was preparing to move to a mobile home park near Portage La Prairie where she had purchased a used trailer. One afternoon I arrived at her house and was surprised when she came running out as soon as she saw me. She was laughing and crying as she explained that she had just received word from the Cancer Clinic that her numbers were right at rock bottom and there would be no more treatment needed. There were celebrations in the dancing crowd that night.

While helping Fran get things ready for her move to Portage, I met her lovely sisters, Marie and Barb and their husbands and many of her friends until it was impossible to keep track of all the new names and faces. Fran was experiencing the same situation as she got to know my neighbours and close friends.

We spent days washing and painting the trailer before friends helped us move her belongings. An auction sale had relieved her of many possessions but it was still a massive undertaking to make everything fit in her new home. The place looked like new and I was starting to get an idea of what a talented lady this mate of mine really was. She could look at an empty wall and picture what it would be like after it was done.

A visit to the Cancer Clinic brought unsettling news. Her cancer numbers were rising again but we were told that it often took awhile for things to settle down after the chemo treatments were stopped. We drove home without saying too much to each other but that nagging worry and sick feeling were back in my stomach and I knew Fran was very concerned as well.

Early October found us in Turner Valley, Alberta, attending a wedding for Fran's friends. We were travelling in my truck and living in an old camper that I had bought that summer.

After the wedding weekend we drove south into Montana and down into Salmon, Idaho to visit some friends of mine, Heather and Lynn Thomas, who I had never actually met in person. We had been long distance friends for years but this was my first opportunity to meet them. They are both tiny people and when we met Heather for the first time, I told her that I hadn't known how small they were. She looked up at us and said, "Well, I never knew that you were giants either." We spent several days camped in their yard enjoying their kind hospitality. I repaid their kindness a bit by helping Lynn cut and haul his winter wood to be piled near their house.

We had made plans to meet Margie and Larry at Emmett, Idaho for a weekend at a cowboy poetry festival so we reluctantly bade our friends farewell and started south. By now some of the wiring in the camper was faulty, the water tank was leaking and the fridge didn't work but we couldn't have cared less. It felt so right and wonderful to be together. We had a roof over our heads and a warm bed to sleep in and cupboards full of food—what more could you ask for? We were acting like teenagers in love and becoming surer all the time that our decision to be together was the right one.

A long day's drive took us to a campground where we shared a happy reunion and a first meeting for Fran with Margie and her hubby. It didn't take Fran long to realise that my sister had the same mischievous devil in her that lived in me. That didn't bother Fran though as she was every bit as much a rascal as I was.

The next day we drove to another campground and set up our two rigs with only one other camper in the park. After supper Margie overheard Fran and I talking about sneaking in to the big building to shower together and so the stage was set for mischief.

With towels in hand, Fran and I walked through the darkness to the shower rooms. We decided to use the men's showers, mainly because I didn't have the nerve to use the women's unit. Once inside, Fran undressed while I had a quick shave. She was just turning on the shower when I realised I needed her help to pull off my new cowboy boots. I sat down on a bench while she straddled my leg and gripped my boot with both hands. I braced my other boot against her bare butt and was ridding myself of the boot when the door partly opened and a man's voice demanded to know, "What's going on in there?" Fran leaped back into the shower as I replied in a shaky voice, "It's just me. I'm having a shower by myself." Once more the voice called, "I said, who's in there?" Again I answered while standing there with one boot half off and frantically trying to get my blue jeans fastened. Fran was staring out at me from the shower stall with her eyes as wide as saucers but a smile on her lips at my predicament. The mystery was solved by the sound of wild, maniacal laughter and my rotten, little sister yelling, "Hey, I got you guys good." It was hard to get to sleep that night because we couldn't stop laughing. This lady of mine, although so very classy and elegant was far from being a prude and she had a sense of humour which matched my own.

The weekend at Emmett was great and was made better when I found out that Margie had entered my name as one of the entertainers. I was able to sing and play some of my favourite songs for the two day show and we met some great people. Fran was a bit embarrassed but also pleased when one man asked me to bring my wife to a seat near the front of the hall.

The month we spent travelling together was wonderful, a time of getting to know each other, talking, laughing and sharing stories as that big, turbo diesel roared a few inches in front of our feet while the endless miles of black-top retreated in our mirrors. Nights were special. With the murmur of a small Idaho brook babbling through the willows just a few feet from our window, we lay awake talking softly in our cozy bed inside the camper. We talked about our former spouses, our kids, our friends and our lives. It was a time of amazing honesty as we confided in each other telling stories of ambitions never realised, dashed dreams, things we had done that we weren't proud of and secret thoughts that we had never shared with another living soul.

One night a fierce storm rolled across the mountains to the little valley where we were sheltered. Jagged lightning streaked across the sky and thunder rolled with the noise of a thousand freight trains as raging winds and torrential rain lashed the camper. That night was the beginning of a total commitment to each other. We promised that no matter what happened in the future, we were in it for the long haul. Tears came as Fran buried her face in my neck and held me close. The storm left the valley and climbed the eastern hills while the tension gradually faded from her body as her mind adjusted to the fact that she wasn't alone any more.

We had left home almost like strangers who cared for each other a lot. We came home a month later as complete partners, soul mates committed to making each other happy for the rest of our lives. We had become husband and wife even though we had no small paper to prove it.

Christmas was shared with our families at my home and her trailer. Her new home looked great all dressed up with some of her own creations. It was our first Christmas together and we welcomed in the New Year by completing plans for a month's vacation in Mexico. I had been told by friends about the little fishing village of Los Ayala, an hour's drive north of Puerto Vallarta.

Fran had been in Mexico several times but it was a first for me. The culture shock was almost too much but I adjusted and grew to love the land and the people. Within days we had met a marvellous young Mexican couple, Jorge and Marbella Nieves who with their two children, Saira and Eric had just moved back to Mexico from Oregon where they had lived for a number of years. This young couple took us under their wing and on their days off from their tiny store, took us touring around the countryside showing us various points of interest. One night we were honoured when we were allowed to take Saira to a fundraising dance in the schoolyard. The three of us danced and ate and made memories to last a lifetime.

One night we heard wonderful guitar music and singing coming from a party in a small courtyard across the street. We took chairs and moved out on the sidewalk in order to see the action. A few minutes later we were spotted and two men invited us to join them. This group from Guadalajara was having a birthday party for one of the men and we were quickly offered cake and tequila. A few of them could speak a bit of English and introductions were made all the way round. I quickly became Daniel, Fran became Francesca and we were welcomed with open arms. I urged them to sing more songs and they obliged with singing the likes of which we had never heard before. I asked for and received permission to bring my video camera to tape the action.

During one song Fran and I started dancing in our bare feet on that rough cement floor. Immediately the tables were pushed back to make room and soon everyone was dancing. Fran and I were taught Mexican dances while the rest continued singing. My partner was a short, stout lady who wore red running shoes. She whirled me and twirled me as she marched me up and down the floor teaching me the various steps while gazing up into my face. My clumsy attempts at following her lead brought great amusement to the group. Fran's partner was a handsome young man who happened to be at least a foot shorter than she was. He danced with great skill and determination with his eyes firmly fixed on Fran's cleavage just a few inches in front of his nose.

I had brought my guitar too and I sang several songs which received a hearty applause. While Fran and I were taking part in all the action, my camera was passed from hand to hand so we were left with a tape of the night. The singing got louder and wilder and soon teenagers started joining the party when they discovered their parents were having more fun than they were. We finally made our escape to our bungalow where we fell asleep listening to more singing across the street. It was an amazing experience and the highlight of our vacation.

We made many new friends among the Canadians down there and some of them will remain friends for life. While walking one evening we spoke to two couples who were sitting in their courtyard and were quickly invited in for a drink with them. Within minutes Rory and Helga Scott from Kelowna, BC and his sister Mary and her husband, Brian Blake from 100 Mile House, BC made us feel as if we had known them for years. It was a sad day in early February when we came home to a frigid Canada.

Back in Manitoba, I dropped Fran off at her trailer and went home to my cold and lonely house. That night my heart was empty without Fran beside me for the first time in ages. The next morning I drove Ruth to the airport so she could start a well earned vacation. Without letting Fran know, I drove straight from the airport to Portage to see my dear one. When she opened the door to my knock, she threw her arms around me and we just stood right there and hugged until the trailer was chilled. One night apart had been much too long and within an hour Fran had packed her bags and we were on our way home to my house vowing never to be apart again.

Within days we were at the Cancer Clinic where Fran went through her usual blood tests. This time the news was very bad as the cancer numbers were exploding again and were now higher than they had ever been. Fran's oncologist explained that the first round of treatment had killed only the easy cancer cells. The resistant cells had survived the treatment and now they were multiplying rapidly. We asked about starting chemo again and the doctor did his best to discourage us, saying that the chemo might knock the cancer down for awhile but as soon as it was stopped, the cancer would grow again. Fran asked him straight out what her odds were if she did take the chemo again and he said he thought she might have a 15% chance of a positive result. Her courage was amazing as she asked him point blank how long she might expect to live if she didn't take the treatment. He told her there was no way of knowing because there were so many variables involved. Fran told him she would think it over and let him know within a few days. We walked out of his office hardly daring to look at each other for fear of breaking down.

That night we lay awake holding each other close but saying little. Inside, I was seething, raging at the whole world, at cancer, and especially at the careless doctor who for months had made light of Fran's complaints about symptoms like bloating, telling her not to worry because she was just getting her middle-aged spread. By the time her cancer was discovered, the damage was done. Her abdomen was filled with five litres of cancerous fluid and a tumour was growing. Now she was paying the price for this carelessness and neglect.

It meant everything to us that we both had our faith in God to carry us through this very difficult time. Fran told me she wasn't afraid to die but she felt resentful that she had to die just when she was totally happy for the first time in her life. As for myself I felt nothing but terror in my heart while imagining life without her.

Three days later Fran told me she wanted to start chemo again, saying that 15% chance was better than no chance at all. I told her I had known what her answer was going to be and had been just waiting for her to say it. An appointment was quickly set up, the chemo started again, and life became as normal as it could under those circumstances. We continued to go dancing almost every weekend but only I knew the effort it took for Fran to arrive at the dances looking elegant and classy, with her make-up perfect and every hair on her head in place. I was the one who saw her struggle to apply make-up with shaking hands while feeling ill and weak. At least she had me to help with her hair. This chemo treatment was somewhat different and she would keep the beautiful hair that had come in curlier than the original. It was a lovely shade of grey and I had convinced her not to colour it as it was truly beautiful the way it was.

It had become hard for her to climb in and out of that old camper so we went shopping for a used fifth wheel camping trailer. It took several weeks of shopping around before we found one that we both loved. We put it to good use travelling to fiddle fests and dances in southern Manitoba and several fishing trips to areas farther north. At times she was too weak to go out in the boat with me but she was always there to meet me at the door with that happy smile and a warm embrace.

There were a series of special gatherings that gave us great pleasure. We had joined with a group of friends to form a small band. I played fiddle and guitar and sang. Mary Riddlell and Ruth played guitar and sang. Phil Leitold played bass guitar and his wife, Rose, played mandolin and guitar and was a great singer. Later we were joined by Harold Backman who played fiddle and guitar and his wife, Adeline, who played guitar and sang. We played for free wherever we were asked and on alternate Wednesdays we played at the care homes in Eriksdale and Lundar. Fran sang right along with us and acted as our sound technician.

It became obvious that she had no need for her trailer at Portage so we decided to sell it and move her belongings to my house in the Interlake. Many long, tiring trips were made to pack up and haul her possessions back home. My house became our house as we gradually found room for everything.

By late summer the news from the cancer clinic wasn't good. The cancer markers had fallen when the treatments resumed but now they were rising rapidly again. Near the end of August Fran's doctor decided to stop the treatments because at this stage they were doing her more harm than good. He offered some pills that could help slow down the growth of the cancer. Once again Fran asked him straight out how much time she had left and this time he told her it might be seven months or less. I looked at my dear one through eyes filled with tears, feeling as if I was caught in a recurring nightmare. The look she gave me spoke volumes but she held her head high and said not a word as we walked out of the building. Back in the privacy of the car, we broke down and cried in each other's arms. The end of the seemingly endless treatments had come; there was nothing more we could do except get her affairs in order.

That night I wrote a letter to our Mexican friend, Jorge, telling him the bad news. They had been expecting us back in Mexico that winter and I wanted to let him know that we wouldn't be coming. Eight days later I got a phone call from Jorge. He told me that if we could get Fran back to Mexico in time, Marbella's family would help us with a natural remedy that Mexicans use to treat ovarian cancer. At first I was sceptical but as I listened to his story I felt excitement rising. Jorge told me there were several women living right in Los Ayala who had beaten the cancer by drinking tea that is brewed from a weed that grows in the pineapple fields.

I told Jorge that we would get back to him and hung up the phone. Fran had been sitting at the table listening to every word and she was beside herself with excitement as I told her what Jorge had said. Within a few minutes we had made up our minds to go for it. We could live in Mexico cheaper than we could at home. The treatment would cost nothing more than the occasional trip to the pineapple fields to pick the weed. We would have a holiday in warm weather and enjoy ourselves. If the treatment didn't work, we hadn't lost anything and if it did work, then we had grabbed the golden ring. There was no hope for her if we did nothing and now, like a miracle, we had hope restored once more.

It was near the end of October by the time Fran managed to sell her trailer and we finished up a few last minute chores to get the house ready to sit empty all winter. On November 11 we boarded a plane and within hours were touching down in Puerto Vallarta, Mexico. Jorge and Marbella were there waiting for us with warm hugs and we were soon on the highway headed to Los Ayala.

That evening Jorge and I drove out to the pineapple fields where the weed grew in profusion. We picked a good supply, roots and all, and headed back to town. Marbella's mother showed us how much weed to use to make a three litre batch of tea. By the next morning Fran was drinking the tea and pronounced it quite pleasant. She had been instructed to drink as much of the tea as possible and in fact it was recommended that she drink nothing but tea for best results. It was made very clear to us that we must also have a strong faith in God and His grace in order for a miracle to happen. We prayed fervently for that miracle.

A few nights later Jorge introduced us to a Mexican lady who had been sent home to die after chemo treatments for ovarian cancer failed to work. She had started drinking this herbal tea and now, many years later she was a picture of health. This reinforced our belief that the decision to come back to Mexico had been the right one.

Life soon settled into a pleasant routine. Many of our friends were back and we were making new friends every day. Fran was feeling stronger now that the chemo was leaving her system. She was doing a fair bit of walking but only in the various towns in the region as her stamina wasn't up to walking the jungle trails that I loved.

We had been trying to learn the name of the weed to see if we could import a supply of it to Canada when we went home. The Mexicans knew it only as Little Bird Plant. One morning while walking on a jungle path near Punta Raza, I met a man who was a professor at a university in Guadalajara. He was with a small group of biologists who were studying the flora and fauna in the area. I told him about Fran and her tea and asked if perhaps he could identify the weed for us. He gave me his email address and said if I could send him a clear picture of the weed, he would see what he could do. I did one better by going straight to the pineapple field to get a fresh supply and then walked back over the mountain path to deliver it to him personally.

Every sunrise would find me heading out on another adventure, never knowing what I might find around the next bend. At times I walked alone until Fran's lifelong fear of snakes caused her to beg me not to go without at least one travelling companion. When I realised how real her fear was, I never walked those trails alone again. Sometimes I would walk as many as fifteen miles before returning to our bungalow, often as late as noon. Fran would hear my key in the door and would dash across the room to greet me with a hug and kiss. Other times I would return to find her gone on another excursion of window shopping. She seldom brought home more than a few groceries or fresh fruit but at times she surprised me with yet another walking companion. I teased her that she had to stop dragging in strangers off the street and sending me alone with them to the jungle.

I ate a late breakfast while she ate an early lunch before walking to the nearby beach where she sunbathed on a mat while I swam for half an hour. Early afternoons were siesta time, drowsing through the soft, humid heat of a Mexico afternoon before waking to begin preparations for supper. We seldom ate out, trying to stretch our budget as far as possible. We didn't need fancy food or the company of others to be happy. We were so in love, so wrapped up in each other, so very tender and tactile; it was as if we weren't happy unless we had almost constant physical contact with each other. We always stayed close enough to be touching the other, sometimes a friendly arm draped across a shoulder, more often a hand lightly touching a thigh, perhaps just fingertips caressing the back of a hand which would cause the other to turn with a smile and an "I love you so very much" look.

The long evenings were special. Fran had insisted that I bring my violin as well as my guitar with us and had even checked in the fiddle case as her second bag. Not an evening went by that we didn't have music. Fran sat quietly with her book of word games and I played our favourite tunes. No music session would be complete until I had played our favourite waltz, Shequandah Bay. More often than not, our only light would be supplied by one or two candles which lit up all except the darkest corners of our bungalow. We could watch small, shifty geckos scurrying across the walls or a procession of tiny ants marching in under the door and across to the kitchen area where they would mill around looking for stray crumbs. This was our wild night life and it was enough to make us happy.

Other times we would take guitar and fiddle and walk to Rory's to share music and singing which often brought out their Mexican neighbours to call for encores. Later we walked home through narrow streets which were almost devoid of lights and felt completely safe. Shadowy figures would appear and pass with a soft, "Buenas noches, amigo." At times like that we felt as if we had come home.

We celebrated Christmas morning holding hands over our tree which happened to be a beautiful poinsettia that we had bought from a peddler. Christmas dinner took place with our neighbours who organised a huge pot luck meal complete with delicious turkey. New Year's Eve was celebrated at El Marlin Restaurant in Guayabitos with new friends, Gilbert and Lorette Rioux from St. Pierre Manitoba.

New Year's Day brought our first argument and hurt feelings to poor Fran. It all started a few weeks earlier when I received an email from Fran's dear friend, Phyllis Bates, saying that she and Garnet were planning on coming to visit us for the first week of January. She wanted me to book a room for them and to keep it as a surprise to Fran. I quickly spoke for a room for them in our hotel. Our hotel host would meet them at the airport but he asked me to ride with him in order to identify them. I didn't want to take a chance on making a slip to Fran so left it until the actual day to tell her that I had been asked to ride with him to Puerto Vallarta to pick up some clients. Unfortunately, Fran had bought some wonderful chicken and potatoes and had invited neighbours in our building for supper that evening. Too late I told her about my trip to the airport and she insisted that I tell the host that I couldn't go. That was the simple thing for me to do but instead, I insisted that she invite our neighbours for lunch or else leave it for another day. Things got rather heated for awhile and Fran couldn't understand why I was being so stubborn. At one point I was on the verge of telling her everything but managed to hold my tongue.

It was two rather angry people who lay down on the bed for our afternoon siesta but we managed a quick hug and kiss before I left for the airport. I asked Fran if she was angry with me and she replied, "Not angry, just hurt." I left there feeling about as low as I could get.

At sunset I arrived home with our guests. Fran was asleep on the bed when we walked in and was totally confused when she looked up and saw Phyllis smiling down on her. Of course, no explanations were needed then.

That was a wonderful week and it was made even better with the arrival of my friends, John and Pat Fjeldsted who drove down from their home in Yuma, Arizona to spend some time with us. That same week we celebrated Three King's Day with our Mexican friends, Jorge and Marbella and their kids. This is the day that all little kids are supposed to receive presents and with the help of friends, we held a party for them in the courtyard of our bungalow. The kids were ecstatic, the parents overwhelmed. Jorge looked at me with his eyes brimming with tears and said, "Oh, Dan, you make me feel so humble." I answered, "No, Jorge, it is you who make us feel humble." This was a young man who was having an extremely hard time trying to make a living but would give the shirt off his back to a stranger on the street.

We had been scheduled to go home in early February but with Fran doing so well we decided to stay longer to give the tea more time to do its work. We changed our flight to March 13, extended our travel insurance, and we were all set for an extra month in paradise.

The month of January brought a steady stream of visitors. Ruth and our friend, Mary came for a week, as did Mickey and Anita from Lundar. With my fiddle and two guitars, we made music which could be heard for blocks. Don and Bev Jarvie from Kinosota spent a day with us and on that same day we had Fran's niece, Cathy and her husband, Mark, from Vancouver Island. Fran hadn't seen Cathy for quite some time and the love between them was obvious to all.

Near the end of January we had a quick visit from my daughters, Donna and Tracy. The sisters had booked a week's vacation in Puerto Vallarta and on January 29, Donna's birthday, they hopped on the bus to La Penita. We met them at the market and treated Donna to a birthday cake at our favourite café before grabbing a cambia to our village. We were so excited to have them meet our Mexican family. By evening they were on their way back to their hotel but it had been a wonderful day.

February brought a change for the worse in Fran's health. She seemed to be growing weaker and was suffering discomfort and pain in her abdomen which she blamed on indigestion and bloating. I was doing most of the cooking and trying to tempt her appetite but nothing appealed to her.

On Valentine's Day, I surprised her with a beautiful pendant made from Rainbow Obsidian. This strange rock can be carved in a certain way to produce a shape deep inside it, a butterfly for example. The one I chose had a perfect heart inside it which I thought was very appropriate for this occasion. It was a big hit and my sweetheart was extremely pleased.

The next few weeks were rather strained with Fran trying to hide her discomfort from me. We spoke little about it as if our silence would make it go away. One morning I arrived home from my walk to find Fran with her gloved hands in the sink washing dishes. I walked straight across to her and put my arms around her from behind while kissing her neck. When my hand touched her lower abdomen she flinched and gasped with pain. At first she tried to make light of it but the time had come for a serious talk. Looking closely at her stomach I could see that the right side was considerably larger than the left. This was the side where the mass was growing. There was nothing for it but to carry on until we arrived back in Manitoba but at least now the situation was in the open and we could talk freely to each other about our fears and her pain.

Early March found us making plans for a double birthday party at Rory's house. Countless singing parties at our good friend's house and return visits to our bungalow had made us grow very fond of these fine people. Fran shared her March 8 birthday with Rory's wife, Helga, and we wanted to make this a wonderful party for our ladies. To add to the excitement, sister Rose and her friend, Peggy, from Prince George, BC were down for a couple of weeks in Guayabitos and they had been invited to the party as well.

The afternoon of the party, Fran seemed very quiet and withdrawn. When I gently asked her what was wrong, she replied that it felt strange to be planning a birthday party that might be her last. There were no words that could help but a long, tender hug helped both of us regain our usual high spirits. It was a wonderful party with dinner out in a restaurant and then cake and fruit followed by music and singing back at Rory's house.

Next morning I received an email from the university professor saying that he had identified our weed as a very common one. It was Green Carpetweed and it grew across the States as well as most provinces in Canada. I made a few phone calls and discovered there was nothing preventing us from bringing a box of it back to Canada. I went to Jorge's place where we had a large supply drying and packed a huge box full to take on the plane. I was quite sure we would have some serious explaining to do at the airport but we were determined to try.

The following day we received a visit from new Canadian friends, John and Judy Soden and John's sisters, Maggie and Jane. I had met Maggie a few days earlier while out walking and we had exchanged stories. When I told her about Fran and the weed tea, she told me that her brother had lung cancer which had spread to his shoulder. Chemotherapy was holding his cancer at bay but he was also taking part in a cancer program in a biomed center in Tijuana, Mexico. We had been urged by other friends to check this clinic out and it seemed that John was getting positive results there. We visited for awhile before I rode with them to the pineapple fields to pick some of the weed so John could try it. It was wonderful to see the fierce love that John's family had for him. I felt like I was leaving old friends when we parted.

On our last day we walked around town saying goodbye to good friends. The last stop was at Marbella's parent's bungalow where we found Jorge and Marbella and some of the other family gathered. By now we knew that Marbella's father had been diagnosed with stomach cancer and the outlook was not good. He and I embraced and called each other Amigo and looked deep in each other's eyes. I saw no fear there, only sad resignation. Last hugs and kisses were for Jorge and Marbella and their kids. We finally pulled ourselves away and walked slowly back to our bungalow with tears falling as we considered this might be the last time we saw our wonderful Mexican family.

Later that evening we discovered our box of weed was crawling with insects of all descriptions. The decision was now out of our hands. There was no chance that we could export the weed with live insects in it and it was too late to go to the fields to pick a fresh batch.

The following afternoon found us back on the farm shovelling deep snow from the path leading to the house. Kind neighbours had used a tractor to clean the yard. The wood furnace was lit, the house became cozy and it was almost as if we had never been gone.

Monday morning we were at the Cancer Clinic in the city where Fran had blood work done before her oncologist checked her out. He seemed surprised to see her looking so well and an internal examination revealed no mass that he could detect. With fresh hope rising in our hearts we drove home to await the results of the blood tests. Next day the results of the test proved much worse than we could've imagined. The cancer numbers had exploded over the past four months. The weed tea hadn't done its work and there would be no miracle.

At breakfast next morning Fran told me she wanted to go to the bio-med center in Tijuana. It broke my heart when she said this could be her last chance and if she didn't go, it would be curtains for her. It didn't take long to make a phone call to the bio-med center and get all the details. Within an hour I had booked flights to San Diego.

On March 31, 2009 we boarded a small plane in Winnipeg, with hope bringing a sparkle to Fran's eyes and a smile to my lips. Our plane had been delayed leaving Chicago and darkness had settled over the land by the time we reached San Diego. The next morning we were up early to catch a shuttle over the border to the clinic. A kind American couple, Jan and Ken, took us under their wings and were a huge help to us. The crossing over the border was painless and we never even had to get out of the van.

At the clinic Fran was taken to another room to have blood tests taken. I was trying to concentrate on reading when I saw several people walk in. It was John and Judy Soden and his sisters. They were on their way home from Los Ayala and had stopped at the clinic so John could have another check-up. They were as amazed to see us as we were to see them and it was so good to touch base with them. It was nearly closing time before we got in to see a doctor. He spent twenty minutes reading us the results of the blood tests and another twenty minutes explaining their program. Ten more minutes and fifteen hundred dollars later we were out of there with instructions to come back for the next batch of tonic in three months.

That evening we had dinner with Jan and Ken in a fancy restaurant. It was encouraging to talk to Jan. After a couple of years in the program, she had now been declared cancer free.

By noon the next day we were on a plane which took us back to our connecting flight in Denver. Shortly before boarding, it was announced that our plane to Winnipeg was overbooked and they were looking for four people to give up their seats for some compensation. I wasn't interested but Fran checked it out and found they would give us each travel credits worth $600 USD, fly us to Chicago and give us a hotel room and meal vouchers. Seeing that we planned on flying back down to the clinic in three months, we felt this was too good to pass up. We were already feeling quite up-beat from the results of the past few days but now we were like teenagers on their first trip to the bright lights. It became even more exciting when our room turned out to be in a large Hilton hotel connected to the airport. Our spirits dimmed a bit when we realised that our travel insurance would run out at midnight. Luckily all went well. We were back in Winnipeg by noon and soon were shopping in a health food store for some of the ingredients for her new diet.

Fran's program called for her to drink a very bitter tasting tonic and go on an extremely restricted diet. I couldn't help but notice that the tonic looked, smelled and tasted exactly like the tonic that our friendly scam artist in Winnipeg had prescribed for Margo. That tonic too, had been shipped from the San Diego area. I kept my doubts to myself, not wanting to disturb Fran when she had new hope.

So began a very trying period as Fran struggled to drink that bitter medicine and we both tried our best to eat the new food and pretend we liked it. Within days, the tonic had made Fran quite ill. I emailed her doctor in Tijuana and explained her problem. He advised us to cut back on the tonic for awhile but stick to the diet as much as possible. She grew to hate her new diet saying that anything that tasted good was off limits. I learned to make bread from unbleached and whole wheat flour but she quickly grew tired of that as well. To add to her miseries, the pain in her abdomen had become almost unbearable to the point where she was taking as many Tylenol as was allowed. Despite her pain and discomfort, she never complained to me. I would often wake in the middle of the night to find her gone. She would be sitting quietly playing solitaire on the computer or playing her word games in the living room. When I asked her why she didn't call me to join her, she said it was important that I get my rest.

By the fourth week of April, Fran was desperately ill but she refused to consider going to the hospital. She had become so weak I was afraid to leave her alone for any length of time. I was busy trying to get my farm machinery in good order for my farm auction which would take place on June 20.

Things came to a head on April 24, when Fran became so ill she started throwing up. She gave in to my demands that we go to the hospital and a quick call to the hospital confirmed there was a doctor on call. Fran showed courage beyond belief as she slowly shuffled step by agonising step out the door and into the car. Every tiny bump in the road made her gasp in pain and it was a great relief to see the lights of town appear.

Fran was kept in hospital for a few days and released on April 27. The following day I drove her to Selkirk for a CT scan. She was given a prescription for stronger pain pills and within a few days she was feeling a lot better. She even started doing the cooking again but was avoiding the tonic which her system just could not tolerate. A real highlight for her was a visit from her son, Todd. Fran hadn't seen Todd for several years and was just over the moon to have her big boy there from Calgary.

The morning of May 3 dawned like any other morning but things were about to take a surprising turn. While working on the machinery, I suddenly became dizzy and nauseous. I staggered up to the house and called Fran to come quickly. She was there in seconds and I told her that I might be having a stroke. She looked at me with panic showing in her face and almost shouted, "No, you're not," as if she was forbidding such a thing to happen. Within minutes I was violently ill, throwing up until there was nothing more to come. Sweat streamed off my face and body in rivers and every time I opened my eyes, the room whirled around me. I was too unsteady on my feet to walk so made several trips to the bathroom on my hands and knees with my eyes closed tightly. By noon Fran had seen enough and called two neighbours to help me to the car. Now the situation was reversed. It was her driving me to the hospital and I was so very thankful I had her looking after me.

My illness was caused by inner ear infection and I spent several days in the hospital. Fran spent hours sitting with me and one evening she and I were lying side by side on my narrow bed while watching television. Doc Burnett walked in and looked shocked as he asked us which one was the patient. I left the hospital using a walker but soon discarded that when I got home.

Ruth had been on a trip to the coast and on May 27, Fran felt strong enough to drive with me to the city to pick her up. It felt like old times to be sitting together, laughing and teasing as we drove. After we collected Ruth we stopped at Barb's house for lunch. Later we shopped for foods which were compatible with her special diet. The tonic had been sitting unused in the refrigerator but we were still following the diet as closely as possible.

Sunday, May 31 was special when we decided at the last minute to attend the afternoon dance in Lundar. Our friends were delighted to see us and we had a great time even though we didn't dance much. Fran was very weak and I was still suffering from vertigo so we settled for a few slow waltzes. We drove home holding hands and talking about more dances to come. We had no way of knowing it but this had been our last dance.

The first week of June found Fran getting weaker with no appetite, and the pills didn't seem to be helping much. On June 8, Ruth was out helping me clean the shop and sort out the tools which would be sold at the sale. Fran was very weak and apologised for not being able to eat much of the supper I had cooked. That evening she had just taken her pain pills when she became violently ill causing her to throw up her medication. Her pain became so intense that she consented to going to the hospital but refused to allow me to phone an ambulance. Luckily, Ruth was still there and between the two of us we managed to walk Fran to the car.

In Emergency, intravenous was started at once and morphine administered to bring her pain under control. It took several bumps of morphine before the pain eased enough to allow x-rays to be taken. She was too weak to stand against the machine so I donned a lead apron and stood there supporting her. The x-rays showed some fluid in her lungs which explained the chest pain she'd been having. At last she was settled into a bed and soon was in a drugged sleep. The next day it was decided to switch to morphine pills which dulled the pain even if it didn't eliminate it.

On June 10, I drove to the city to keep an appointment with an ear, nose and throat specialist. The specialist did several tests and found that I was suffering from two different types of vertigo. There was little that could be done but she said that if I kept active, the vertigo would gradually lessen.

Back in the country I drove straight to the hospital to find Ruth and Mary there serenading Fran. There had been other visitors as well and Fran was happy to see them all. Troy arrived to spend time with his mum, sleeping right outside the hospital in the old camper I had sold him. Over the next few days he was an immense help to me as we got things ready for the sale. We took turns sitting beside Fran at night and those nights I spent there will always be precious in my mind. We teased a bit and giggled but also had serious talks as we shared each other's pain. Fran told me one night that she hoped I wouldn't date anyone for at least one year after she died. I told her truthfully that I didn't want to date anyone ever again. My sweetheart was wise and said that a time would come when I would want to date again and she only wanted me to be happy. All the while we had been together she had a way of making her eyes go big and round when I may have been feeling down about something. Now she made her big eyes at me again and I laughed and hugged her.

On the morning of June 14, Fran was very weak and needed help to walk from the bed to the bathroom. That evening Todd and his girlfriend arrived from Calgary. Fran was overjoyed to have her strong, handsome sons sitting beside her and gazed at them as if she had just won the lottery. I sat with Fran that night and was there next morning when she was moved into the palliative care suite in the hospital. It felt strange to see her lying in the same room where Margo had died and my heart felt so heavy it was hard to keep from crying. At the same time, it was wonderful for her to have privacy and more room for visitors.

Troy finally ordered me to go home and get some sleep and I was happy to oblige. I was sound asleep that afternoon when Troy phoned me in a panic and asked me to come as quickly as possible because his mother wasn't expected to last the night. Even though I knew Fran wasn't doing well, it was a huge shock to hear this. I broke speed records all the way to town and prayed to God to let her live long enough for me to see her again.

At the hospital I found Fran just barely awake. It was her turn to be surprised when I threw my arms around her and smothered her with kisses. She wanted to know what was going on and I told her I had missed her a whole lot. Over the next hour this extraordinary lady amazed all of us when she insisted on getting out of bed and sitting in a chair visiting with the boys and me. She was a bit mixed up from the medication but otherwise just like her old self. The nurse suggested we take her out in a wheelchair for a breath of fresh air. Fran thought that was a great idea but insisted she had to put on makeup first. Knowing how important it was to her to look her best, I suggested to the boys that they go for supper to give her time to get ready.

The next hour was both hilarious and sad as my sweetie carefully put on her face. Her little mirror wouldn't stay propped up so I sat in front of her and held it as she worked. At times she would almost fall asleep until I reminded her that we had places to go. This was the first time I had actually watched the whole process and it was rather fascinating. At one point I must have been watching too closely because she suddenly put down her little brush and demanded, "And just what are you staring at?" I told her I was just looking at her because I loved her. The boys came back and started wheeling their mum out of the hospital while I ran down to the store and picked up a little disposable camera to take some pictures. The lilac bushes on the corner were blooming and we picked a sweet smelling bouquet to take back to her room. It was a special evening seeing how happy she was with her boys by her side. I took lots of photos but as I write this, those pictures are still inside my camera. Some day soon I will have them developed and the tears will fall again.

The next few days were hectic doing last minute jobs to be ready for the sale. At times like that you realise how important it is to have good friends. Garnett helped me move machinery into place and fix flat tires while Phyllis stayed with Fran in the hospital. Einar came from Woodlands to help do some mechanical work on engines. Todd spent hours washing the combine. He did such a thorough job, the fuel tank started leaking and Einar and I had to remove the tank and cold weld it. My sister, Dot was there from Bellingham but was staying with Ruth. There was precious little time to spend with my dear old sister.

All spring Fran's sisters had been very faithful at driving out to the farm to visit. The day before the sale, they sorted through the huge pile of Fran's belongings. Fran was very upset at the thought of parting with any of her favourite possessions and had given them careful instructions about what to sell and what to keep. That afternoon I had lots of help as we cleaned out the shop and steel shed and loaded everything on huge flat rack trailers. With that job done we started on the household stuff. It looked as if it might rain so we moved the loaded trailer into the steel shed for the night. The ladies were played out but went around by Eriksdale to see Fran before going home. An hour later they were back with fresh instructions. Fran had changed her mind about selling a lot of her things and we had to go out to the shed and sort through it all again. The keep pile in the back corner of the shed was almost as big as the sell pile on the rack by the time we were done.

June 20 dawned bright and clear as I'd been praying it would. My farm auction was scheduled to start at 10 AM sharp. By eight o'clock the faithful workers from Central Plains Cancer Care were setting up shop in the entrance to my large steel shed. These dedicated people work hours on their own time to raise funds for their organization and today they would provide coffee and lunch for what we hoped would be a large, hungry crowd.

People started arriving and the sale got started right on time with Buddy Bergner calling for bids on the household items, most of which belonged to Fran. I spent the morning trying to avoid talking to anyone, even old friends. It was a horrible feeling to watch a lifetime of hard work being auctioned off. The sale ended in the late afternoon with the auctioning of my beautiful '93 Dodge truck which had carried Fran and I in our home away from home all the way to Idaho. With the sale over, there was time to sit on the shady deck and share a beer with Fran's sons and their ladies and my daughters. Troy and his family soon headed home to Portage while Todd went to the hospital to spend the night with Fran.

By June 23 Fran was having a rough time. The morphine was making her agitated and very demanding as she invented little chores for me to do around the suite. It didn't bother me as I had been through this before and knew it would take awhile for her to adjust to her new medication. That night Mary stayed with her so I could get some much needed sleep.

June 25 was better. Mary, Ruth and I played music for Fran and later her old dancing partner, Herb, showed up. We persuaded him to play and sing several songs for Fran and she was very happy. She was much stronger and that evening I took her out of the hospital for a short walk. Next morning Fran was thought well enough to go home again. All her medications would be contained in blister packs so I would have no trouble being her nurse. It was amazing to be heading for home with Fran beside me, beaming with happiness, when we had been so close to losing her the week before.

The next few days were difficult and we struggled to find the right amount of medication which would keep her pain at a tolerable level without making her too drowsy. Beside the blister pack meds, she was allowed short acting pills which controlled the breakaway pain. It helped immensely when her friend, Jean, came to spend a night and a couple of days with us.

July 1 was a special day with a visit from Phyllis and Garnett. Our band had been asked to play on the street in Eriksdale for the Canada Day Celebrations. Our friends offered to stay with Fran so I could take part but Fran had other ideas. She said she would very much like to go along with us and watch us perform in front of the crowd. We had an early supper and headed for town to get set up. Fran and our friends set up lawn chairs across the street from our stage on top of a large flat deck trailer. When I played Fran's favourite waltz, I could feel her love flowing across the street to me. She was in a lot of pain but stuck it out right to the end so we could drive back home together. We held hands all the way.

Two days later I made a quick trip to Lundar to pick up stronger morphine pills for Fran. Her pain at times was unbearable. Her abdomen was becoming distended as the mass inside her grew larger. I had a secret fear that I kept between me and God. I had heard about two ladies whose ovarian cancer ended up with the cancer eating its way right out through the wall of the abdomen. I prayed to God to please not let this happen to Fran.

On July 4 Elmer and Linda Nickel drove down from Steep Rock for a short visit. They had never met Fran but they all seemed to like each other. We started telling them about our winter in Mexico and the weed tea. Fran told them she wished she was back in Mexico because she had felt so well while drinking the tea and now she was feeling so crappy. Next morning

Fran offered to pay my fare if I would fly back to Mexico and bring back a large amount of the weed so she could start drinking the tea again. I thought she was joking but one look at her face told me she was deadly serious. I told her I wouldn't be allowed to bring it and she told me that I knew darn well that it was perfectly legal. I said that if I tried to get it through the airport in Mexico, I would likely find myself in jail. It was hard to look at her as she faced me with tears in her eyes and her voice pleading. She said, "If I don't get some help soon, it will be the end for me." I felt as low as I could get as I told her there was no kind way to say it except that the tea hadn't worked. She made one more comment about not having given it enough time and then the matter was dropped, never to be mentioned again. I felt as if I had abandoned her in her greatest hour of need and that familiar, helpless feeling was back as if it had never left.

On July 9 Fran was admitted to the hospital. Her doctor was not happy with the amount of fast acting pills Fran had been taking for breakthrough pain even though she was still within the limits he had prescribed. He planned on trying her with the pain patch and keeping her hospitalised for several days to monitor her condition. His plan worked well and within a few days she was feeling better than she had for a long time. On July 13 she was released from the hospital. At home I took her straight to my vegetable garden so she could admire the flowers I had planted for her that spring.

The roller coaster continued for the next few days. Fran had become very emotional to the point where she couldn't stop crying. For the first time she didn't want to see any visitors or talk on the phone. On July 16 we got visitors anyway when Phyllis and Garnet dropped in to stay a few days. Fran was delighted to see them and her mood brightened at once. Next day Garnet and I went fishing and left the gals to amuse each other.

The next ten days were quite good. Fran was eating little and was becoming alarmingly thin and quite weak but she was so cheerful and undemanding that each day was a blessing, a perfect gift from God. We had plenty of time to talk about the future, hers as well as mine. We both believed that she would be going to Heaven and we often talked about what it would be like there. Would she get to see Fred and her parents and maybe even Margo? I asked her to tell Margo that I loved her and missed her and Fran promised she would. I teased her about whether she and Margo would somehow be allowed to have their large selection of shoes there and be called the Imelda Marcos twins.

Fran was terribly concerned about leaving me behind to fend for myself. She wanted me to go back to Mexico to walk in the sun and mend my heart. I told her I couldn't stand the thought of being there without her. She said it would be good for me to see our Mexican family and other friends.

Now that she had resigned herself to the inevitable, she started making plans for her own Celebration of Life Service. It would be held in the hall in Kinosota, just down the road from where Fred was buried. She sat at her beautiful kitchen table with shaking hands carefully making out a list of things to do. Daily the list grew longer. She wanted to be cremated and as we had previously discussed, her ashes were to be split. Half would be buried with Fred and the other half would remain with me to be joined with half of my ashes when the time should come. She chose her niece, Terri, to do her eulogy and Phyllis' daughter, Debbie, to sing three songs. She even wanted my cousin Bill and I to sing Amazing Grace. The service would be followed by internment at the cemetery, then lunch at the hall. To top it off, she wanted our band to play for a little dance for our friends following the lunch and I promised to do my best to arrange that.

July 26 was a special day. Fran had told me early in our relationship that she had two half sisters. I had met the older sister, Marge, several times but on this day Fran would meet Debra, her younger sister for the first time. Barb and Marie and their hubbies arrived after lunch bringing Debra with them. Fran had been beside herself with excitement all morning and now here they were together at last hugging and talking, getting to know each other. It was a lovely day and we all moved out on the deck in the shade of the maple tree. I took photos of the bunch and of Fran and Debra together. Later we enjoyed a wonderful meal the sisters had brought.

The next day Fran threw a curve that just about floored me. Her niece, Darcy, and nephew, Perry, had been married on the same year and now they and their mates were teaming up to hold a combined 25th wedding anniversary. Fran told me she would like to attend the party at Reg and Darcy's yard near Kinosota that weekend. This idea was coming from a very sick lady who could barely walk around the house. She said "We'll take the truck and the fifth wheel camper and if I get tired I can just go to bed. If I get really sick, I'll stay in bed and you can drive me back to the hospital." Knowing how much this meant to her, I told her I would get the camper ready and if she felt well enough, we would go.

On July 29, my son Steven and his family arrived from Alberta for a visit. They only knew Fran from talking to her the previous winter on Skype but warm hugs were exchanged. The kids were shy with me and I made a silent vow to see more of them. They soon left to take part in a family reunion for Linda's side of the family.

On the morning of July 31, a man arrived to pick up the crimper for the swather he had bought at my auction. Promising to be back quickly, I left Fran playing cards on the computer. I loaded the crimper in his truck, and then hurried back in. Entering the house I could hear her calling me. I found her lying on the floor of the office, unable to rise. Shortly after I had gone out, the phone rang and she had jumped up to answer it. She had stumbled over her own feet and fallen heavily against the sideboard before crashing to the floor. She was too weak to regain her footing and had to stay right there until I came back fifteen minutes later. She had large bruises on her shoulder and one hip but there were no bones broken. She took it as a matter of course, saying it was her own fault for being so clumsy, but I could see the incident had shaken her up.

To make matters worse we were expecting company; Robert and Jackie Isfeld were stopping in for coffee and some music on their way back to The Pas from the city. We had also asked Jackie's brother, Cameron, to join us. I wanted to make a phone call and cancel our date but Fran wouldn't hear of it. That evening she sat in pain trying to look as if she was enjoying herself. I knew she was doing this out of her love for me and my heart went out to this courageous lady.

Next day we drove to Kinosota to take part in the anniversary celebrations. Our hosts were expecting us and when we pulled into the driveway, Reg waved at us to park right beside the house. It was a beautiful day and the guests were seated in lawn chairs. We set up chairs in the shade of our camper and within minutes Fran was playing the part of queen for the day. Most of the folks were strangers to me and after awhile my head was whirling trying to keep track of them all. Everyone knew how ill Fran really was and I could see the sorrow in their eyes as they looked at their dear friend.

A huge buffet meal was served and Fran insisted on loading her own plate with the delicious food. After a few bites she pushed her plate away, unable to eat. Later I joined a group of musicians and we made some very good music. Darkness was falling when we said goodbye and drove over to Phyllis and Garnet's cottage a few miles away. Fran was tired but also very happy when we settled down in our comfortable bed in the trailer.

Next morning we had a leisurely breakfast with our friends in the cottage before driving south to visit more of Fran's neighbours. As we approached Fran's former house where she had lived a good part of her life, she asked me to slow down. I asked her if she wanted me to pull into the driveway but she told me to just keep going. She was pleased to see the improvements that the new owner had made.

That night we camped just a mile further down the road at Jack and Gloria Robertson's place. Jack was in the city but their son, Clint, and his mom made us very welcome, serving us a beautiful supper. More friends stopped by including Jim and Crystal Bruce and Fran's dear friend, Hazel. Fran asked me to bring my guitar from the trailer and sing some songs for them.

Finally it was time for hugs and kisses and as they drove away, I helped Fran walk across the yard to our trailer. She was barely inside the door when she broke down and wept bitterly while I cradled her in my arms. I didn't have to ask what was wrong. This was the last time she would see her friends. What was going through her head I will never know but within minutes she had pulled herself together and was thanking me for bringing her.

We had breakfast the next morning before driving a few miles to visit two more close neighbours, Marge and Bill Minton. They insisted that we stay for lunch and Fran felt well enough to eat. After lunch we headed for home. By the time we passed the Lake Manitoba Narrows, Fran was becoming extremely tired. Reaching Hayland Hall, I pulled in under some huge oak trees and we slept for an hour. The day was growing late when I woke but Fran was still exhausted. Just as we had planned, I left her in bed and drove home. We had done what I had considered the impossible. It had been a marvellous weekend and had given Fran a chance to say goodbye to countless friends and family members.

Two days later, Mary came to be with Fran while I took my grandson, Kevin, fishing up at Anama Bay on Lake Winnipeg. That was a marvellous day bonding with a grandson who admitted that his memories of our farm were fading. Since they moved to Alberta we had seen little of each other so this day was extra sweet to me. It pleased me immensely to see how his enthusiasm and patience for fishing matched mine. By late afternoon we were on our way home with our limits of pickerel.

Another day I made good progress at clearing a camping spot on a piece of my land on the lake shore. It felt good to work until the sweat ran freely after so many weeks cooped up in the house. It was also a good spot to sit and talk to God and ask Him to watch over us in our trials.

Friday, August 7 was nothing short of wonderful. Fran was acting just like her old self, even doing some house cleaning as she trotted from room to room. At one point she gave me a tight hug and told me that if things kept going this way, she might beat the darn cancer yet. That afternoon we had a lady stop by to arrange some home care so I could get out of the house once in awhile. Fran hadn't wanted any outside help but she could see that I was becoming burned out.

If Friday was good, the next day was as bad as it could get. Fran was weak and withdrawn, unable to eat and quite confused. The weekend passed and things slowly got worse until I finally suggested we go back to the hospital. She rejected that suggestion at once, saying she just needed time to gather her strength. She thought she might be able to eat some raspberries so dear Ruth quickly made a trip out and brought us some. Fran tried some with yogurt and managed to get them down.

On Monday, August 10, Fran was much worse. I was so relieved to see Marie and Barb drive in. I spent a few hours with tractor and brush mower knocking down weeds and tall grass near the barn. The sisters made a nice supper but Fran was unable to eat. Her pain was reaching a point where the medications had no effect. About ten o'clock she threw up. I knew from past experience that with her medication gone, we had only a short while before she would be overcome with pain.

It was time to go but still she refused to leave, saying that she would be alright after resting awhile. I told her very firmly that if she didn't make an effort to move I would phone the ambulance to come and pick her up. She then agreed to go but only after she had a complete wash. She was too weak to stand in the shower but I was able to give her a good wash while she was sitting in the bathroom. Then I wanted to take her straight out to the car but she said first she had to go back to the bedroom.

She sat on the edge of the bed and slowly turned her head to take in all her lovely possessions, her bedroom suite and pictures, twin lights above the bed and even the curtains on the windows. Finally she turned to me and said, "Okay, I guess I'm ready to go."

By then she was so weak, it took the three of us to get her moved out to the car. The sisters sat in the back seat while I drove carefully over the bumpy country roads. Six miles from home I reached the pavement and put the pedal down. Fran suddenly turned and put her soft hand on my arm and said, "We sure crammed a lot of good times into two years, didn't we?" The light from the dash lit up her beautiful smile and for an instant it was as if we were just on our way to another of our beloved dances. This time we both knew there would be no returning home together.

It was close to midnight by the time we had her settled in the palliative care suite. Her sisters took my car back to my house and I spent the night in the chair beside Fran. Once the pain was under control, she slept.

Next morning the gals were back to sit with Fran while I went home to catch some sleep. I returned in the afternoon bringing a large bouquet and Fran's eye's lit up when she saw them. I put down the bouquet and we hugged and smooched until our visitors were becoming embarrassed. The sisters soon left for the city, promising to return early the next morning.

It had become very difficult for Fran to swallow her pills and it was decided to control her pain through the intravenous line which had been started when she arrived. Her upper body was thin and frail but her legs and feet were badly swollen which was upsetting for her. All her life she had been plagued by fluid retention which had been controlled by medication but now there was nothing which could be done to prevent it.

On Wednesday, August 12, the sisters returned as promised. Fran continued to have visitors and was totally gracious and loving to all of them no matter how ill she was feeling. I decided to go to Lundar with our band to play for the old folks in the care home. It was a release for me and our good friends to be able to make music which brought much pleasure to those old people.

Back at the hospital I found Fran quite confused but she brightened up when Rose Leotold and I played music for her for half an hour. Once again her sisters left for home promising to return the following afternoon. That was a long strange night as I sat with Fran. She seemed to sleep little, turning her head to look out the window and then looking back at me. Every time I spoke to her she would look at me and smile but didn't speak. About five in the morning, our good nurse, Werner, came in to check her condition. Every time he asked her a question, she just gave him a frightened look. He realised at once that she was unable to speak. He was so gentle with her as he told her she could nod or shake her head yes or no to his questions. This she was able to do easily which made it easier for him to see to her comfort.

About 6 AM she finally fell asleep. I walked to the kitchen and told the cook not to bother bringing any breakfast for her but asked for a small dish of apple sauce which she loved. When the tray arrived I got her propped up in bed and started feeding her. He mouth opened eagerly for each spoonful and she had no trouble swallowing. I asked her if it was good and she nodded her head up and down. "Not too sweet for you?" She shook her head for "No" and smiled at me. My heart was breaking but I was trying not to cry in front of her. I had never been able to fool her though and she knew exactly how I was feeling. Trying to be brave, I gave her a big phoney smile and she gave me her big-eyed look and a crooked smile right back.

After breakfast the nurses came to wash Fran and change her bedding. I was sitting in the living room when a nurse came to tell me I should call her family because they didn't think she would last much longer. Unable to use the room phone for long distance, I called Ruth and asked her to call Marie.

The nurses were with Fran a long time and when I was allowed back in the room, I found her in a coma. There would be no more smiles between us. She was very restless and in pain and just couldn't seem to settle down.

By ten-thirty I could stand it no longer and phoned Rose to see if she would help me play music for Fran. She was there in minutes and soon the room was filled with the music that Fran loved. At Rose's suggestion we played only slow waltzes. That old fiddle and guitar had never sounded better as we played our hearts out for the dear gal we loved so much. With each tune, Fran became more relaxed and finally was lying quietly with a little smile on her face. I played her favourite, Shequandah Bay, and then ended with Amazing Grace. When the last note died away, I told Rose it would be alright now and thanked her so much for coming. She bent and hugged Fran before she left. We went into the other room and had a little cry and a big hug before she left.

Drawing a chair close to Fran's side, I held her hand and talked to her. There was no response until I softly sang one of her favourite hymns, "I'll Fly Away" without throwing in the extra, "in the morning" because I knew she didn't like that. The song finished, I whispered to her, "It's time for you to fly away now, Fran." Her eyelids flickered and her hand gently squeezed mine so I knew she had heard everything.

Then it was just a gentle relaxation as her breathing slowed. Twenty minutes later, she drew one last soft breath and sighed just like my dear Margo had done. Under my hand, her heart slowed and then stopped. She was gone, her beautiful spirit leaving her tired body and flying away free from pain for the first time in years.

On August 18 we held one last party for Fran. Her Celebration of Life Service went exactly as she had planned. The hall was overflowing with our friends and the service was perfect. The wonderful minister, Jona Weitzel, hadn't missed a single detail. The sun shone down on the little cemetery at Kinosota which had been trimmed up neat as a pin by kind neighbours. Half of her ashes were laid to rest with Fred.

Lunch in the hall was followed by a dance and we ended it with a rousing rendition of I'll Fly Away just to make sure she made it all the way to Heaven. She had requested no tears that day, just happy memories but of course that was impossible. We smiled through our tears and sang for one very courageous lady.

The next month was a blur of tears as reality set in. At first there was the huge relief that she was no longer suffering, then the sad realisation that she really was gone.

By the end of September, I had decided to honour one more of her last requests. I had resisted the idea of returning to Mexico for the winter. Now I realised that it was something I must do, if for no other reason than to find closure and to write this last chapter. On this January day as I sit here in my bungalow in Los Ayala looking out the window at palm trees swaying in the warm ocean breeze, I feel her presence with me. I can almost hear her laughter in the wind as I walk my favourite jungle paths. This winter she walks with me and can see for the first time all the wonders I had only been able to tell her about.

Some day, if God chooses, I will see Fran again and Margo too. Until then, I will hold you both in my heart. Hasta Luego, my dear ones.

Dan and Fran in Mexico, January 2009.